MacNamara Russell

Mount Lyell Mines, Tasmania

MacNamara Russell

Mount Lyell Mines, Tasmania

ISBN/EAN: 9783744790406

Printed in Europe, USA, Canada, Australia, Japan

Cover: Foto ©berggeist007 / pixelio.de

More available books at **www.hansebooks.com**

MOUNT LYELL MINES,

TASMANIA.

———————•◦•◦•———————

Note.—This Book contains a few up‑to‑date Particulars of the above Mines, and forms an Appendix to the writer's Book "Mount Lyell Mines," published in October, 1896.

EDITED AND COMPILED BY

MACNAMARA RUSSELL, M.Inst.C.E.

(WITH SMALL LOCALITY MAP SHOWING THE PRINCIPAL MINES.)

PRICE FIVE SHILLINGS.

EFFINGHAM WILSON, ROYAL EXCHANGE.
1898.

INDEX.

Scale 2640 Feet to 1 Inch.

0   2640   5280   7920 FEET

Compiled by M. Russell M.Inst. C.E

LOCALITY PLAN SHOWING POSITION OF THE PRINCIPAL
COPPER & GOLD MINES AT MOUNT LYELL
AT PRESENT DATE.

QUEEN MINING Co.

LYELL PROSr ASSn

WESTERN COMSTOCK MINING Co

TASMAN LYELL MINING Co

CROWN LYELL MINING Co

LYELL COMBINED COPPER LIMITED

MT LYELL BLOCKS MINING Co NL

WEST MOUNT LYELL COPPER MINING Co NL

LYELL FARRELL MINING Co NL

LYELL CONSOLS MINING Co NL

ROYAL MINING ASSn N.L.

WEST COMSTOCK LYELL COPPER MINES OF FED LIMITED

PRINCE LYELL

COMSTOCK MINING ASSn

STORE CK MT LYELL MINING ASSn

EAST MT LYELL MINING Co NO LIA

M & R Co I.P

MT LYELL

GLEN LYELL MINING Co NL

CENTRAL MT LYELL MINING Co

LYELL CONSOLS GOLD & COPPER NL

MT LYELL MINING Co

MT LYELL PROPRIETARY LIMITED

LINDA LYELL MINING Co N.L.

FREDER MT LYELL MINING ASSn N.L.

MT LYELL MINING & RAILWAY Co LP

SOUTH MOUNT LYELL CONSOLS Lp

SOUTH MT LYELL MINING Co

MT LYELL MINING & RAILWAY Co LP

MOUNT LYELL PROPRIETARY LIMITED

TOWN OF GORMANSTON

GREAT SOUTH LYELL MINING COMPANY

MOUNT LYELL MINES,

TASMANIA.

INTRODUCTORY.

My Book, "*Mount Lyell Mines,*" *Tasmania,* was so liberally bought by the Public—and has been so uniformly praised both by the Press and by a very large number of voluntary and impartial testifiers to the substantial advantages which have accrued to themselves and clients, &c., by their having followed the information data and advice contained in that work—that I am encouraged to believe the present small "Supplement" or Appendix thereto will receive its proportionate degree of support and approval, and that it will also serve the other purposes I have in view : (i.) the strengthening and consolidating the high opinion which the "Big" Lyells have evidently so abundantly obtained amongst a large section of our shrewdest and wealthiest financial and other men : and (ii.) that my previous readers may now find in the pages of this present small Supplementary publication the admission and assurance that owing to the very many rapid developments which have taken place in connection with several of the surrounding and adjoining mines (and notably so in the "North Lyell"), my conservative statements in the Preface to my Book—as to my not then venturing to believe that the "outside Mines" had a rival amongst them to the Big Mine itself—now require to be considerably modified in view of the almost phenomenally rich and extensive discoveries which have been made in "The North

Mount Lyell Mine"; and in view also of the numerous new
"shows" and "iron outcrops" &c. which are being vigorously
prospected and opened up by quite a host of more or less well
capitalised Companies in the Lyell field.

Many other very important changes and conditions of affairs
and all of them as I consider highly favourable in their bearing
upon the future success of this field, have occurred during the
past eighteen months or so; and therefore in the few following
pages I have endeavoured to crystallize into crispest form possible
an outline of the present (and probable future) condition and
prospects of these Mines.

Very many of these Lyell Mountain ranges, spurs and valleys,
and probably a considerable portion of the very extensive area
forming the slopes of "Mount Owen" &c. will in my opinion be
found to contain payable bodies of ore amongst the vast beds of
cemented "conglomerates" which form hereabouts so conspicuous
and promising a matrix. I myself found gold—in minute particles—in these "conglomerate" rocks taken at hazard from the
disintegrated fragments thereof forming the bed of "Conglomerate
Creek," and as it is now known that quite a large number of
"iron outcrops" have been recently discovered located and prospected in the immediate vicinity of the Big Mine I cannot refuse
to believe or to admit that "Lyell" may now be fairly regarded
as no longer merely a "one-mine" field, such as Mount Morgan,
Mount Bischoff, or Rio Tinto and Tharsis, &c., but that it bids
very fair indeed to produce quite a large number of mines having
large and rich Copper ore deposits; and whether these are "lodes"
proper or merely isolated "deposits," and whether of aqueous or
of plutonic origin matters very little if there be payable ore in
sufficient abundance; for instance, as in the case of the doubly-
lucky "North Mount Lyell," with its two large and rich "East"
and "West" Iron outcrops and "lodes" or deposits and also a
centre ore body of large size and value.

Whilst then freely admitting that practically a new order of
things in many respects has grown up at Lyell since I first
ventured to place my "interested" views before my readers some

eighteen months since I still adhere—in stubborn conservative fashion—to the opinion I then expressed in my Book, viz.: that even yet no absolute "rival" to the Big Mine has been *proved* to exist; but I am "liberal" enough to qualify this expression of opinion to the extent indicated in the preceding paragraphs.

"Au reste" . . . well, it is already "plain as a pikestaff" (to the initiated at any rate) that two very potent factors have been in active operation—(i.) a "Bear" attack on the Big Lyells themselves; (ii.) a fairly numerous crop of flotations of properties more or less dependent—as their only claim for existing even on paper—to the now magical name of "Lyell," and that for some time to come there will be produced for the delectation of the generous and short-memoried B.P. the usual percentage of Wild-Cat Mines "goes without saying."

I also however now feel certain that not even the brilliantly fortunate North Lyell Mine is to be the only other successful or possible rival which the "Big" Mine is to have amongst its neighbours; and that whilst the "Big fish" such as "Mount Lyells" "North Lyells" &c. will soon be looked upon as "rich men's" stock, and as partaking more of the nature of "Investment" Stocks than of "Speculative Shares," there will be at least a few fairly safe "gambles," and some of them of really very promising and even important character, amongst the more immediately adjacent and surrounding mines; and of these Lyell Mining Stars of lesser magnitude (so far as is yet known) I have in the following few pages given as full—yet concise and impartial—particulars as are procurable by me at the moment of going to press.

And I have no hesitation in recommending my apparently fairly numerous mining and Exchange friends to remember that not even a few years' success in market operations—in such shares as those of the "Big" Lyell, *par exemple*—will ever yield to the "operators" or "short-profits" and "quick-returns" market-men, the solid and substantial fortune-making profits which the operator for "long-view" results can generally safely reckon upon. Of course, the latter class of operator in addition to being a "long-

view " individual must be able to "finance " his views ; but as an
instance of what I mean let me say that scarce a day or week
passes but I am told by one or another of my Exchange and
other speculative friends . . . . "what a fool I was not to
" buy Lyells (in 1894) when you told me to do so at 27s. or there-
" abouts, and quietly locked them up in my safe and kept them
" there, instead of dabbling in and out for a quick turn or two,
" and at last getting afraid to touch them when they are at
" high figures."

Or yet another—sorrowing like Job—" Just think of it, . . .
" say I had bought 1,000 Lyells in 1894 at 30s.—and kept them—
" my £1,500 of invested capital would yield me a profit of, say,
" £15,000—or at the rate of 1,000 per cent. . . . or, if
" reckoned as an 'income' producer, would in dividends yield
" me at least **£1,500 per annum—(or cent. per cent. per
" annum)."**

As a now many-years member of and operator on Stock
Exchanges, of course I am familiar with this sort of "lamenta-
tion "—which, whilst being common enough, is after all but a
species of "jobbing backwards," and as futile as fretting about
spilt milk, &c.; but the "moral" which it is in my mind to
draw from these "opportunities missed" may be summed up by
pointing out to some of my Exchange friends that it is sometimes pure wisdom in mining speculations as in other of our
worldly affairs "to despise not the day of small things or
beginnings," and that very frequently indeed solid fortune-
making is achieved by "him who waits," and who allows time
for a mine to be developed, or the seedling to become a tree.

The natural-born ' Bears ' of the Market, who are "**in**" everything to-day and "**in**" nothing long, or not longer than they
can cut a loss or score a small turn of profit, will not I know take
the risks of " long-views " or " mine-views "; and certainly a life
spent on 'Change is well calculated to undermine any man's belief
in men, mines and mundane matters generally. Yet, even to
such pessimists " of little faith," I venture to commend the Big
Lyells as being still " good buying."

In my Book in October 1896 I pronounced them to be at their then see-saw prices of say round about £9—" **The Greatest Potentials of the Mining Market;** "—and I **consider them now still more justly entitled to be thus regarded.**

At the present time I again unreservedly admit I am a less "disinterested" adviser however than ever, seeing that practically my own chief Mining ventures—either as Mine owner or speculator—are those at Mount Lyell.

But in the past this "interested" condition of my position has not proved very disastrous, I may safely claim, to those who have admittedly paid me the compliment of having acted upon the information which I have placed before them, and I therefore reiterate the head-lines which heralded my first pamphlet (by an "Interested" Party).

> . . . . "To be taken, then, '*cum grano salis*'.' . . . .
> By all means; but also 'without prejudice.'"

M. R.

3 GREAT WINCHESTER STREET, E.C.

THE

MOUNT LYELL MINING & RAILWAY CO.

LIMITED.

"THE BIG MINE."

WHY ITS SHARES ARE STILL "GOOD BUYING."

Nearly as regularly as "clock-work" the Smelters have been
familiarising the lucky Shareholders, and the Public, with
four-weekly returns of Copper Gold and Silver from ore of
"average" yet of unprecedented high grade, and of the most perfect
quality for cheapness and facility of Smelting hitherto known to
Copper Mining—and Copper Smelting—professional men; in fact,
in both these very important respects, the Big Mine's average
ores are vastly more than fulfilling Dr. Peters' best expectations
and predictions, and have established—or "broken"—the record
in such matters.

Other big and world-known historic Copper Mines have, we
know, had vaster deposits than those as yet developed at Lyell;
but not one of these many gigantic—and in many cases most
magnificently "payable" Mines of great and deserved repute—
can "hold a candle" to the Big Lyell Mine in these matters of
"high grade" in Copper, in Gold, and in Silver combined; and all
these wombed in a matrix unique in respect to the complete
absence therein of one particle or element of a "refractory"

character ; " an ideal ore," in fact, as it has been aptly and justly, as well as unanimously, proclaimed to be by the many competent judges who have examined and tested it : and now, foremost amongst such men must of course be prominently placed Mr. R. Sticht, the Company's Chief Manager, upon whose well-merited promotion to which position the Shareholders should indeed hugely congratulate themselves.

Under almost unparalleled difficulties of one sort and another, Mr. Sticht has already " done wonders " in bringing the present " half-plant " of five Smelters progressively into blast, and in keeping them " going " with comparatively only trivial breaks or shut-down intervals.

And Shareholders would do well to bear in mind that even under all the disadvantages referred to, these Smelters have been already earning "profits" or "dividends" largely in excess of Dr. Peters' estimate. And that when—(during A.D. 1898, as we are informed by the Directors in their Report will almost certainly be the case)—the whole ten Smelter Plant is in full blast, there should be a 'dividend producing,' or net profit over and above all working expenses whatsoever, sufficient to admit of from £2 to £3 per Share per annum being paid upon this Company's £3 Shares.

Dr. Peters asserted in 1893 that at any rate a net profit of £1 per ton of ore treated could be safely depended upon ; and on the assumption that 1,000 tons of ore would be treated per diem for say 300 days in the year, we find that £300,000 would even under Dr. Peters' modest computation be available for distribution on the 250,000 Shares of £3 each which have been issued to the Shareholders (out of the 300,000 Shares of £3 each forming its Capital).

At the present moment this Company's fortunate Shareholders are confronted with the following unchallenged and unchallengeable and highly pleasing prospects :—

(i.) That the result of the past half-year's operations has given a ' net profit ' of £66,724, notwithstanding the fact

that the whole five Furnaces were not all "in blast"—
or running—until the end of the half-year.

(ii.) That the cost of producing Blister Copper during
this period has been reduced to **£1 3s. 3d.** per ton of ore
(as against Dr. Peters' estimate of **£1 16s. 0d.** per ton
of ore.

(iii.) That the ore was estimated by Dr. Peters to contain
copper 4·5 per ton of ore ; silver 3 ozs. ; gold 2·5. The
actual average grade of the ore smelted up to the present
date (January 1898) has been proved to be as nearly as
possible of the above estimated values as regards their
respective percentages ; and their average money values are
as follows :—

		Per Ton of Ore.		
Copper 4·5 ; valued at £50 per ton	...	£2	5	0
Silver 3 ozs., „ 2s. 3d. per oz.		0	6	9
Gold, 2 dwts. (about), at £4 per oz.		0	8	0
		£2	19	9

which is at present slightly against Shareholders as compared with Dr. Peters' estimate in 1893.

(iv.) A strong "Reserve Fund" has been inaugurated.

(v.) **The surplus of liquid assets over liabilities
is £177,812 at date of November Meeting 1897.**

(vi.) **The Shareholders have been officially informed by their Directors that no further issue of
Shares or Debentures will be required, there being
sufficient money in hand to pay for the completion
of the full "10-Smelter Plant," &c. ; so that there
will be actually only 250,000 Shares "issued" (in
lieu of 300,000 as contemplated originally).**

(vii.) That, on the principle of "two strings to a bow"
being generally considered better than one, the policy of
the North Lyell Company in providing an additional
railway line to Port Macquarie, and the deepening of the

entrance to its waters &c. is essentially a wise one; and at the same time there cannot I consider reasonably be any objection on the part of Shareholders in the Big Company to their Directors either individually or in a representative capacity promoting or encouraging another and distinct "Northern" outlet to the Emu Bay waters.

Rumour has had it that this "Emu Bay" part of the Big Company's policy includes the providing presently of a sufficiently abundant means for the treatment of 5,000 tons of ore per diem at Emu Bay, in addition to the local smelting of 1,000 tons per day at the Penghana (Lyell) Works; but the Chairman repudiated all such "wild talk" about 40, 50, 60 Smelters &c. &c.; so there is apparently no official authority for the statement. But, all the same, there are plenty of advocates who are in favour of treating Lyell's "life" as one to be 'shortened' to the utmost, by erecting even 50 or 60 Smelters—seeing that there is ore enough "in sight" to then require 5 years to work it out at rate of 5,000 tons per day.

There will also to all present appearances be an enormous output of ore from some of the surrounding and as yet "silent" Mines such as "The Copper Mines of Mount Lyell West," "Central," "South Mount Lyell," "Mount Lyell Consols," "Mount Lyell Extended," "Lyell Tharsis," "King Lyell," "Mount Lyell Proprietary," and probably from many others both north and south of these Mines; and therefore the more railways and "outlets" for the vast requirements of such properties the better for them.*

(viii.) So gigantic a forecast as is here indicated could only be justified by the certainty that so huge a quantity of ore as say 2,000,000 (two millions) tons per annum will be forthcoming; and as various "Bear" estimates have recently

* On December 15 1897 *The Mount Lyell Standard* published the following par. :—"The surveyors for the Emu Bay Railway Company have examined the country from the junction of the existing line to Zeehan and Lyell, and have reported favourably upon available routes to both places. Arrangements have been made for the location of further survey parties, who will work out from Zeehan and Lyell towards the other parties. The Company mean business."—M.R.

been passed round " the House "—and elsewhere—fixing the
visible quantity of ore now in sight at the Big Mine (down
to its present deepest level of 500 ft. below Crotty's " Iron
Blow ") as being far less than the real developments appear
to justify, whilst certain other " Bullish " estimates of such
" ore in sight " have overstated this " visible " quantity, it
will be worth my reader's while to work out and prove—or
disprove—this quite simple arithmetical question for themselves ; and the following data and sectional sketch diagrams
are furnished herewith to enable such calculation to be made
—and verified.

THE AMOUNT OF ORE " IN SIGHT "
IN THE BIG MINE.

Like unto nearly all other arithmetical problems, this question
of the visible amount of ore supply in the Big Mine entirely
depends upon the accuracy of your premises, or of your postulates
constants &c., &c. ; otherwise of course " figures " are capable of
producing the most misleading and yet apparently perfectly correct
results.

As a " constant " then whereupon to base our estimate I
invite my readers to believe that the conservative Dr. Peters in
his earliest full report upon this Mine (see page 66 of my book
Lyell Mines), in fixing the " constant " at about 8 cubic feet of
ore to the ton of 2,240 lbs. fixed it too high ; my own numerous
experiments for ascertaining its specific gravity having afforded
as a fair " mean " constant some 7·0 cubic feet of ore per ton.*

* I would regard 6·5 cubic feet as being still nearer the true constant.— M.R.

PLAN SHOWING CONTOUR OF PYRITES ORE DEPOSIT

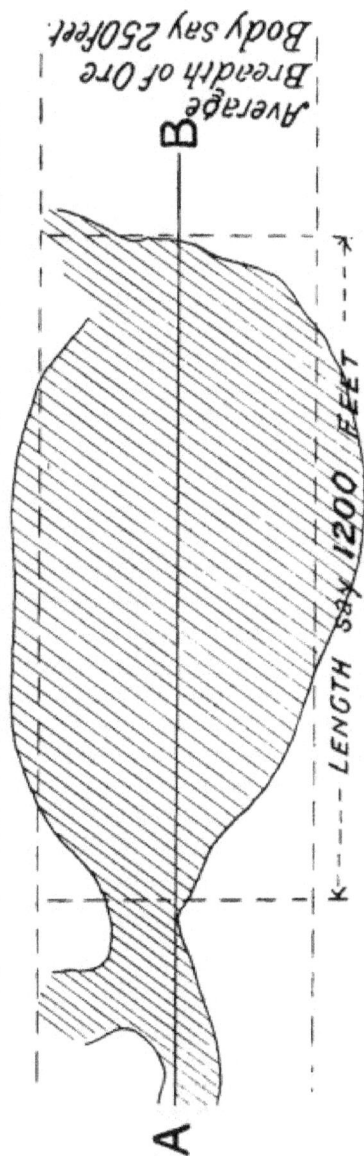

Average Breadth of Ore Body say 250 feet.

B

A

LENGTH say 1200 FEET

The following " vertical plane " section through this " deposit " or lode on about line A, B, affords further data for our calculations and estimate of the measurements of the ore body already exposed down to the No. 5 Tunnel Level (1,000 ft. above sea level).

Average length of ore body disclosed to date say 1,200 lineal feet.*

Average width of ore body between its

hanging and foot-walls   ...   ...   say   250 lineal feet.

Average depth of ore body from surface

down to No. 5 Tunnel Level   ...   say   450 lineal feet.

Constant for weight = say 7·0 cubic feet per ton of ore, and thus

1,200 ft. × 250 ft. × 450 ft. = 135 millions of cubic feet of ore,

and $\dfrac{\text{Cubic feet.}}{135,000,000}$

Constant 7·0   = 19 millions of tons of ore in sight.

This ' visible supply ' of ore would be sufficient to meet the ever-craving demands of Smelters " in blast " as follows :—

			Term of Year's supply.
10 Smelters at 1,000 tons per day for 300 days per annum			63 years.
20 Smelters at 2,000	do.	do.	31½ ,,
40 Smelters at 4,000	do.	do.	15¾ ,,
60 Smelters at 6,000	do.	do.	7½ ,,

But magnificent as is this " prospective " life or—endurance— of the Big Mine as disclosed by this bit of simple arithmetic, it falls far short in various ways of its probable potentialities, since these really comprise many other ascertained and further " prospects," viz. :—

(a) There are **known** " zones " of rich ore bodies in the various levels of the main-lode deposit which must add immensely to and raise the average of the lower grade ores.

Note : * Dr. Peters in his Report of 15th May 1893 estimated the then known length of the "lode" at 800 feet ; but some hundreds of feet of additional north and south "Drives" &c. have been put in during the past four years—and these justify apparently—our now estimating the total length of exploited "ore in sight" at about 1,200 feet.—M.R.

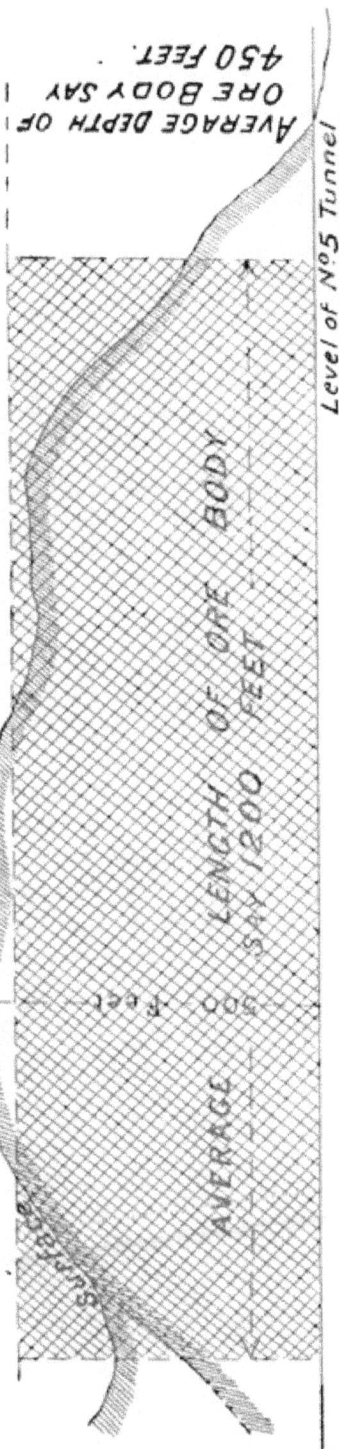

SECTION ON A VERTICAL PLANE THRO' THE ORE BODY
on Line A B.

Crotty's Iron Blow

AVERAGE DEPTH OF
ORE BODY SAY
450 FEET.

Level of Nº 5 Tunnel

LENGTH OF ORE BODY
SAY 1200 FEET

AVERAGE 300 FEET

Surface

(b) The Mine can be worked as a "dry" Mine for probably 150 ft., or 200 ft. below the present 500 level, and the "19 millions" of tons would be thereby increased by one-half = say 27 millions of tons before the question of "Pumping" would even require to be thought of.

(c) And each additional 100 ft. of depth should *pro rata* give a further 4 or 5 million tons of ore.

The question then of "supply and demand," or "life" of this Mine, does not admit apparently of a moment's doubt; and it is quite clear that whether that excellent expert Mr. Sticht did or did not "calculate" that there were "twenty-two millions of tons of ore in sight," the Mine itself is seemingly its own best and irrefutable "authority" on the point. What matters it to us whether Mr. Sticht's reputed post-prandial estimate of "22½ millions in sight" was made or not?. The figures used by me in my calculations above given can be "discounted"—to almost any extent which may make the most pessimistic unbeliever happy—and still leave "Big Lyell" in the unquestionable position of having the biggest visible supply of ore yet seen in any Mine at its nativity.

Cut down—discount—my Estimate by one half and the " 10 smelter " plant can still safely rely upon being supplied with its daily 1,000 tons of ore per day for say 30 years. Whilst 60 smelters—with their daily wants of 60,000 tons of ore—could be kept going for 3½ years.

Is not this "good enough" to satisfy even the most exacting of pessimists?; I think so.

WHY BIG LYELLS HAVE DROOPED LATELY?

(Query: "Ursa Major" *versus* "Southern Cross"?)

"Why?" indeed :

Certainly not for any just or reasonable cause whatsoever, so far as the Mine itself is concerned, or so far as the Smelting Returns are concerned, for both these "vitals" are as full of "vitality" of the most satisfactory character as is well possible.

Many of my readers must have noticed (as I have) how it happens pretty frequently now in these days of thirty-five days' trips to Antipodean shores, that there is quite a swarm of visitors from Capel Court to the "down under" regions of "Crux Australis." Lately certain of these visitors have been unmistakably of the "Ursa Major" order of beings and have favoured "Lyells" especially with their very best forms of "squeeze".

Result: timid holders in Melbourne and London have enabled Messrs. Ursa Major and Minor to knock off a couple of "points" or so from the premier "Lyell" stock as a sort of "first blood."

Poor timid sellers! Shrewd clever "Bears."

POTENTIALITIES OF "LYELLS."

Not quite, as yet, have the "Big Lyell" shares become firmly established in financial men's confidence as being undoubtedly the very soundest and safest gilt-edged investment Mining Stock on the World's lists of such goods. But before this present year of Grace upon which we have just entered is ended, I venture to assert that Lyells will have attained that position, **and that they will also have advanced many "points" in market value.**

" Why should they do so ? "

Answer : Let these figures speak for themselves : —

With only the present 5 Smelters running for fairly "full time," their net earnings for dividend purposes during the ensuing half-year **should** be **at least £100,000 ;** and by the end of that half-year **3** additional Smelters should be "running," and should add, *pro rata*, 50 per cent. to the June-December's (1898) Profits ; or say a grand total of £250,000 **at least** available for dividends on 250,000 shares during the next 12 months.

From January 1899 to December 1899 this "full" plant of 10 Smelters with their treatment of 1,000 tons of ore smelted per day at £2 per ton "profit" should earn **at least** £600,000 net "profit" for dividends—or at the rate of **48 shillings per share per annum** or **80 per cent. on each £3 share.**

With such an undoubted length of "life" as has been already practically "guaranteed" by the Mine itself surely then it is not predicting or expecting too much that "Lyells" should steadily

and soon—start again on their up grade. If we assume that 10 per cent. is a fair enough minimum return to be looked for from a Mining Stock then it is as "clear as daylight" (Australian daylight—not of course London "fog" daylight) that a £3 share earning say 48 shillings dividends annually, **is a gilt-edged security of the most distinguished order ; and that on this 10 per cent. basis it would be still a very excellent invest-ment at a market price of £24 per Share.**

Romantically large or absurd as such figures may now seem to conservatively careful investment experts—or to those who will not fairly and squarely " work out " the correct dimensions and " constants " &c. of the Big Lyell lode deposit—they still fall absurdly short of the true potentialities of this veritably mar-vellous Mine. For instance :

A gradual increase of Smelting Plant and Smelters will assuredly form a " constant " part of Big Lyell's progressive policy ; and even a 5,000 tons per day Smelting programme should not be regarded as either a chimerical or an impossible goal.

If then my views as above faithfully carefully and analytically put forward are correct, or even largely and substantially so, what other "listed" or "unlisted" stock is there which can compare for an instant with "Lyells"?.

Serious political complications (" Mail-ed fists "—made in Germany &c. &c.) might of course give even " Lyells " a bad time ; but this contingency is happily one not requiring serious present consideration.

THE ADJACENT MINES.

"THE NORTH MOUNT LYELL MINE."

AT this moment I am face to face with the fact that I can no longer justify—even to myself—the non-rendering unto Cæsar the things which are Cæsar's ; and with such an undeniably very 'Cæsar' of a Mine indeed as this North Mount Lyell property has in its as yet almost incipient or adolescent days demonstrated itself indubitably to be, I will hasten to purge myself of the imputation which I find has sometimes found voice—amongst the readers of my Book—that I did not at the time it was published —(October 1896—) believe in the existence of a possible or probable Rival to the "Big" Mine itself amongst its numerous neighbouring Mining Confrères.

As far back however as that memorable August of 1891 when Lyell was visited by myself—as a member of Mr. Bowes Kelly's Syndicate party—I find by reference to my Field-book containing

the notes of the Surveys I then made of the Big Mine itself as
well as of the more important surrounding blocks—that I also
made notes and sketches of the position &c. of the immense iron
outcrops visible in these "North Lyell" blocks; and that an
entry was also made at the same time by me in that field-book
to the effect **that it was the next "biggest"**
show to "Crotty's Iron Blow."

So I can honestly claim that I am not a mere prophet 'after
the event'—but that I was—even in those now almost remote
historical "Lyell" days—duly impressed with the rugged moun-
tainous assertive and powerful looking iron outcrops which
towered up so conspicuously though they were then deeply
embosomed in the dense 'Bush' or forest which made it so
difficult a task even to such a Herculean mining pioneer as
my friend Mr. James Crotty himself to reach this precipitous
spur—of some 2,000 feet above sea level—of Mount Lyell's
main Peak.

To a more modified extent I was also impressed by the lesser
Iron outcrops (and Copper pyrites deposits) which I found to
exist on certain other "forfeited" Gold leases (seven in number
of 10 acres each), and which my then partner and myself
decided forthwith to re-peg and apply for; they having been
originally pegged out and held by Mr. James Crotty and some
of his party. And these 7 blocks now form—with another 10
acre block more recently acquired—the property of my recently
re-constructed "West Mount Lyell Company"—and now known
as "The Copper Mines of Mount Lyell West Limited."

It is we know quite usual for a parent—or for parents—to
consider his or their own family—or some of its individual mem-
bers—as almost unquestionably far and away the superiors of
'other people's' broods. And as every Mining Man, or Mine
Owner, almost invariably pronounces (—and sometimes possibly
even believes) his own particular mine to be "**THE** very best
one of them all" . . . . it would surely have been merely
natural and quite *de rigueur*—according to the ethics of the game—
if in my Book aforesaid I had sought to place my own "West

Lyell " bantling as the most probable " Under-study " of—or Rival to—the " Big " Mine itself. But this I have never as yet dreamed of doing ; although naturally perhaps now that I find my own Mine amply provided with Working Capital for fully and properly exploring and developing it—I am secretly not without hope of yet seeing it announce itself in similar " undeniable " fashion to its " North Lyell " neighbour—and as fully worthy to even rank as the real " No. 2 " Mine of the " Lyell " group ; a position which—on paper at any rate—it assuredly holds as regards its being ' Big' Lyell's **nearest** neighbour.

But—' Che Sara Sara ' is assuredly about the fittest of mottoes for a Mine or Mine Owner ; although that other still more familiarly known ' couplet ' fits the case almost like a glove also :

. . . . . " For if ' **she** ' will—' **she** ' will,
You may depend on't--
And if ' **she** ' won't—' **she** ' won't--
And there's an end on't—"

These Mining " She's " **are** built just like that—and there's no denying it.

Meantime let us return to " Cæsar" North Lyell and his evidently most abundant Imperial purples blues and greens and gold and iridescent glories of " Peacock " ore of the very finest quality, and in almost fabulous abundance as to quantity.

As though to make up to Mr. James Crotty for his many very weary and at times heart-breaking—as well as back-breaking— years of loneliness, semi-starvation—and deprivation of almost every comfort material and intellectual—and as a sort of ' Con-solation-stakes ' to him for the loss involved in his almost compulsory (certainly never **his voluntary**) surrender of his " Big " Crotty's Blow " Mine under the exigencies of evil Fortunes, the Good Genii of the Mines of Lyell decided that as Mr. Crotty was practically principal owner—and " Lord and Master " of several of the Northern Group of Mines—as well as of the big South Mount Lyell Mine—it would only be ' a dacent thing ' to at last reward this Lion-hearted patient Lyell pioneer and modestly heroic-devout believer in Lyell Mines and in the virtue of patience.

And thus—almost like the gorgeous transformation scene of a Drury Lane Pantomime—the erstwhile frowning, discouraging, toilsome heart and back-breaking "North Lyell" Mine has suddenly thrown off its armoured opposition to this invading plodding and determined Mining Man ("Cæsar" aforesaid) and disclosed to him in one brilliant flash of the Mine-Genii's most dazzling light—the incalculable Treasures of one of his very choicest of Mine-wealth secrets.

I am glad to say that I am amongst those who have known Mr. Crotty for some years past—and no one who is really cognisant of his career on the Lyell field can fail to offer him—as I do—the warmest congratulations upon his Phœnix-like emergence from the toilsome past into the brilliant present with all its almost dazzling possibilities for himself—and for those also who are wise enough to pin their faith to him—and to his advice and counsels as to this Lyell field generally.

For most unquestionably he **IS** justly entitled as an authority to quite the 'first-place' in respect to all that there is to be known about these Mines at this time; and in his other as yet less favorably known properties (such as the "Lyell Extended" and "Lyell Consols" Mines) I am now perfectly convinced he holds "potentials" of such high promise as will presently entitle them to rank if not quite on an equality with the Cæsar-like "North Lyell" still as very 'considerable personages' in Mine circles of quite high degree.

Thanks to the Press mainly and to some few persistent individual "Lyell" enthusiasts—much is now generally known to the English Public about Mr. Crotty's various Lyell properties—and especially at the present moment about his "North" Mine—and about his admirably conceived and admirably carried out projected "Railway Bill"—with its now-attendant immediate "construction" of the line itself &c.—that it would be a mere waste of my readers' time for me to invite them to read any more of my mere personal views and opinions; and having—with every good-will—endeavoured to adequately express my own perfect faith both in Mr. James Crotty himself—in his Mines—and his views and opinions, I will proceed to such work as is before me in referring to other of the less known—and less developed Mines and shows of this Field.*

"THE QUEEN LYELL MINING COMPANY, N.L."

This Mine is situated on the northern descent of the "Lyell West Peak;" *i.e.* an altitude of rugged mountain range—with the culminating West Lyell Peak (say 4,500 feet above sea level)—has to be surmounted—as it 'intervenes'—between the "North Mount Lyell" Mine and the Queen Lyell property—which latter is probably therefore some 3,000 feet above sea level and about two miles north of the North Lyell Mine. I have no personal knowledge to impart about this property never having visited its particularly 'difficult' neighbourhood.

So far as its present mining developments are concerned then I must be content to quote the following extracts from Mr. Lawson's excellent little booklet upon these Mines (published in Melbourne recently).

---

* That well-known Civil Engineer, Napier Bell, of New Zealand, has just paid the Lyell Mines a visit, and he is stated by *The Lyell Mining Standard* to have replied as follows to the questions asked by his interviewers as to the proposed works for deepening and permanently improving the bar and entrance to Macquarie Harbour. . . . "During his stay, which extended over Monday, he was interviewed by a *Standard* representative. Asked if there would be any great engineering difficulty to surmount in undertaking the deepening of the bar, he replied that everything was so simple that the work would not interest engineers. He was convinced that the harbour could be improved permanently, so that vessels of deep draft could steam right up to the wharf at Strahan without the slightest danger. The work that had been accomplished at Greymouth, in New Zealand, in the face of the angry southern ocean, was very much more difficult than the deepening of the Macquarie Harbour."

"EXTRACTS FROM MR. G. A. LAWSON'S BOOK

OF THE

MINERAL INDUSTRY OF TASMANIA."

"THE QUEEN LYELL," N.L.

Capital £100,000 in 100,000 shares of £1 each ; 50,000 paid
up and 50,000 contributing ; (12,500 paid up and 12,500
contributing are held in reserve for the Company.

Legal Manager : G. A. LAWSON, 90 Queen Street, Melbourne

"This property consists of 72 acres held under lease from the
Tasmanian Government. The lode is about 130 feet wide. Its
composition closely resembles other lodes in the near vicinity, and
is located upon what is now known as the Western Lode, which
traverses the West Mount Lyell, the Prince Lyell Company's
sections, and others intervening and not named. It appears to be
of a massive and permanent character. The assays by A. E.
Glover, of 397 Little Collins Street, are very satisfactory, viz. :—
Gold, 8 dwts. 14 grs. ; silver, 42 ozs. 9 dwts. ; copper, 36 per cent.
per ton. This is a very encouraging result, because the destruction
of its metalliferous ingredients must have been enormous in ages
gone by."

"The workings comprise adit and some three or four surface
cuttings across the outcrop. The same facilities exist here for
developing these mines as on others likewise located on the almost
precipitous eastern ranges from the Queen River, which permits,
after once the adits are driven, of expeditious and economic
mining operations. The work already done demonstrates the
present and prospective great value of this lode. So far, at
present, only the south portion of this mine is being developed.

"Ores appear to occur in large bands or laminations, and they
consist of hematites, iron and copper pyrites, and some galenites.
Of these, the copper pyrites are associated with a large quartz

formation. The ores have a close resemblance, and are identical
with those so generally prevailing in this district, and also those
raised by the parent and other companies. The quantity of ore
is large, so far as can be seen at present, and will be materially
increased at deeper levels. It is a property that can be opened
up quickly and operated on at moderate expense."

———————

Having started with about the most northerly of the
"Registered" Mining Company's blocks *i.e.* **"The Queen"**
—(*vide* the small locality plan facing title page of this booklet of
mine) I propose to work down 'Southerly' therefrom and to deal
with the various properties seriatim.

Starting then—(in imagination only I am thankful to say—for
as a pedestrian excursion the 'climb up' Lyell West Peak and the
'climb down' therefrom is a very creditable bit of mountaineer-
ing) from Queen Lyell's 'distant' mine and Regal charms—and
descending Lyell West Peak's main Spur—or "Divide of the King
and Queen River Watersheds"—southerly, we begin to get into
the 'thick' of the Mines: and in about the following order :—

1. Queen Lyell.
2. Crown Lyell Mining Co. N.L.
3. Tasman Lyell Prospecting Association.
4. **North Mount Lyell Limited.**
5. Mount Lyell Blocks Mining Co. N.L.
6. **Lyell Tharsis Mining Co. N.L.**
7. **Mount Lyell Consols M. Co. N.L.**
7A. **Mount Lyell Extended Mining Co. N.L.**
8. West Mount Lyell Consols N.L.
9. West Lyell Extended Mining Co. N.L.
10. "Royal Mining Association" N.L.
11. **Prince Lyell M. Co. N.L.**
12. Lyell Pioneers N.L.
13. Lyell Pioneers Consolidated N.L.
14. North King Lyell G. M. Co. Ld.
15. East Mount Lyell M. Co. N.L.

16. Kaiser Lyell Mining Association N.L.

17. **Central Mount Lyell Mining Co. N.L.**

18. Glen Lyell M. Co. N.L.

19. King Lyell G. M. Co. N.L.

20. **Mount Lyell Proprietary Co. Ld.** (late "Minerals" Block, 40 Acres).

21. Linda Lyell Mining Co. N.L.

22. **Mount Lyell Mining & Ry. Co. Ld. Principal Blocks ("Crotty's Iron Blow") and Main Lode Workings.**

23. **The Copper Mines of Mount Lyell West Limited.**

24. **South Mount Lyell Mining Co. Ld.**

25. **Mount Lyell Proprietary Co. Ld.** (South Blocks).

26. Great South Lyell Mining Co. N.L.

27. South Mount Lyell Consols Ld.

28. Great Southern Mount Lyell Syndicate N.L.

CROWN LYELL MINING COMPANY.

No Liability, Tasmania.

Capital £125,000 in 125,000 shares of £1 each ; 62,500 of which are fully paid up ; 62,500 are contributing.

Legal Manager : G. A. Lawson, 90 Queen Street, Melbourne.

" This block of metalliferous land, 30 acres in extent, is situated north and north-west of the North Mount Lyell Company's lease, and forms an oblong parallelogram, thus commanding a greater length of the lode opened than would otherwise have been the case, and is directly north of Tharsis Company's, also west of the Tasmania-Lyell Company, from the former of which the lode has been traced right into this ground, which is now being opened up with satisfactory results.   Occupying the position it does, this mine is very centrally situated among the other numerous lease-holds on the field, being situated on the north-western slope of the range which forms the watershed, as between the creeks emptying south into the Linda Valley on the north falling into the upper tributaries of the Queen River.   Besides the main lode there are indications of the occurrence of other metalliferous formations."

"The lode worked by the North Mount Lyell Company adjoining has been traced by means of massive outcrops of manganiferous iron into this property. The ores comprise the usual sulphurets of the Mount Lyell mineral field, a combination of iron, copper and silver lead, and as characteristic of this locality become richer on greater depth being reached. The property can be economically worked by means of adits. The north-western strike of the lode exhibits a western underlay, which has been traced through this lease by means of "iron blows." The angle of strike is such as will add considerably to the contents of the ore-bearing formation on this property. Assays have been made, returning very satisfactory results."

(Above Article is extracted from Mr. Lawson's Book.) M. R.

THE TASMAN LYELL PROSPECTING ASSOCIATION.

No Liability.

Capital £15,000 in 30,000 shares of 10s. each; 28,600 shares are issued as paid up to 5s.

Legal Manager: John Potts, 31 Queen Street, Melbourne.

"Operations on this property have been mostly on the surface for the past six months. A low grade pyrites body of 150 ft. in width has been met with giving assays of copper from 1 to 2 per cent., and gold from 1 to 2 dwts. The western side of this formation seems to carry the most copper."

"Another discovery is also being operated on—a pyrites formation about 3 ft. wide, and increasing in width as it goes down, exposing a fair-looking show, about same as other formation. A shaft has been sunk 16 ft., about 6 ft. in nice-looking iron, from which colours of gold can be obtained."

"This property is well worth further consideration, for it has splendid facilities for tunnelling, &c., and the prospects of the association are very encouraging, being in good hands."

(Above Article is extracted from Mr. Lawson's Book). M. R.

THE NORTH MOUNT LYELL COPPER COMPANY,

LIMITED.

CAPITAL - - - - £500,000

Divided into 500,000 Shares of £1 each.

Of these 395,000 Shares will be taken by the Vendors in full payment of the purchase money, leaving 105,000 Shares available for the Working Capital of the Company. Of these, 45,000 Shares have been applied for by the Directors and their friends and will be allotted in full, 35,000 Shares are reserved for future Issue, and the remaining

25,000 Shares are available for allotment to the public.

Payable 5 - on Application, 5 - on Allotment, 5 - two months after Allotment, and 5 - four months after Allotment.

The Shares, when fully paid, may be converted into Share Warrants to Bearer.

Directors.

JAMES CROTTY, 138, Leadenhall Street, E.C., and Melbourne.
WILLIAM JACKS, D.L., J.P., Glasgow, & 23, Leadenhall Street, E.C.
LEONARD R. HIGGINS, 9, Drapers' Gardens, E.C.
JOHN S. MACARTHUR, Glasgow, and 56, New Broad Street, E.C.
D. J. MACKAY, 138, Leadenhall Street, E.C.

Melbourne Board.

ALFRED CLAYTON, C.E. GEORGE MOORE, M.D.
JAMES P. LONERGAN. CHARLES E. PACKER.

Bankers.

London—THE COMMERCIAL BANK OF SCOTLAND, Ltd.
62, Lombard Street, E.C.
Australia—BANK OF AUSTRALASIA, 4, Threadneedle Street, E.C.

Brokers.

DERENBURG & CO., 9, Drapers' Gardens, E.C.

Solicitors.

RENSHAW, KEKEWICH & SMITH, 2, Suffolk Lane, E.C.
LYNCH & MACDONALD, Oxford Chambers, Bourke Street, Melbourne.

Auditors.

SINGLETON, FABIAN & CO., Chartered Accountants, 34, Nicholas Lane, E.C.

Secretary *(pro tem.)* and Offices.

D. G. LUMSDEN, 138, LEADENHALL STREET, E.C.

It will probably interest and repay my readers to glance at the following extract (from Mr. Lawson's book) respecting this property before proceeding to my own further remarks on this phenomenally rich Mine :

NORTH MOUNT LYELL MINING COMPANY, LIMITED.

Capital £150,000 in 150,000 shares of £1 each, 60,000 being paid up and 60,000 paid to 15s. ; 30,000 shares held in reserve.

"The area consists of 30 acres held under gold lease (which embraces all other minerals), and is situated on the southern slope of Mount Lyell and the eastern slope of the ridge between Mounts Lyell and Owen. The general fall of the ground being from the west to the east, it will be seen that the property is adapted for working by adit levels, which is of great importance for economical mining. At about the centre of this property is a formation of barztic hematite and quartzite which marks the course of the metalliferous ore belt underlying to the west. This hematite and quartzite outcrop usually lies on a conglomerate deposit which forms the footwall of the ore bodies. The rock outcrop to the west of this hematite lode is an ironstained schist, and is the locale where pyrites ore veins will be found. Therefore the whole of this property is so positioned that the lode is contained well within it. On section 29/90 an outcrop of dense copper pyrites (contained in schist) has been trenched on in various places, and to prospect this a tunnel was started on section 23/90 to intersect the ore body. At 200 feet in from opening an ore body, "erubescite" of high grade, corresponding with that on surface, was passed through, its strike being east of north. This important development occurring in soft ground has not been further explored. A south drive off western crosscut commencing 82 feet from end of main tunnel, has been following a body of copper pyrites and erubescite. This drive follows the footwall of the ore formation for about 70 feet and is most likely a portion of the main ore body marked on surface by the hematite.

" A winze being put down in the ore body near the footwall
is opening up well, showing a marked improvement in every foot
of sinking, some veins being very rich. The value and appearance of the ore to-day at this level is highly satisfactory, being
only 120 feet beneath the natural surface."

" As this mine is situated so that fairly deep prospecting can
be got by adit, it may be expected that large bodies of payable
ore will be met with as the mine is opened up. The underground
workings already prove existence of payable ore at greater depth.
From evidence of immediate surroundings and workings of other
mines it suggests considerable improvement in size and quality."

" The latest assays made 62 per cent. copper, 30 ozs. 15 dwts.
silver, 15 dwts. gold per ton."

" The Mining Manager has just reported a new find, a lode
having been cut, consisting of galena, in the tunnel. It looks
strong. There is no footwall visible yet."

Mr. Crotty's Railway Bill is thus described in the columns of
the " *Lyell* Standard " on December 11th 1897 and should prove
of interest to many of my readers :—

. . . . . .

"THE NORTH MOUNT LYELL RAILWAY."

" The second reading of the Mount Lyell and Macquarie
Harbour Railway Bill was moved in the House of Assembly on
Wednesday by the Hon. D. C. Urquhart, and was agreed to."

" The total estimated cost of the railway is £92,486 10s. ; the
length, 30 miles 41 chains ; the average cost per mile therefore
will be £3,031."

" The surveyors of the trial line were Messrs. W. F. Egan and
W. M'Eachern, who submitted the following general report to
the Parliamentary Select Committee " :—

" LOCATION.—For 3½ miles from the North Lyell Mine the line
passes down the northern side of the Linda Valley along the
foot-hills of Mount Lyell. At 3½ miles it crosses the Linda

River and the overland track from Hobart to Strahan. This, which is the only road crossed on the entire route, it is proposed to cross by an overhead bridge. From 4 miles to 17 miles the route is through level button-grass plains, devoid of timber, the only works of any consequence necessary being bridges across the King and Governor Rivers. After leaving the plains the line curves round the head waters of the Andrew River, ascending to a low gap in the divide between the Andrew and Bird Rivers. From this gap it descends along the Nora and Bird Rivers to Kelly's Basin, and after skirting the shore of the basin for a mile crosses the Fysh River, and terminates at the proposed wharfage casements on the northern shore of Kelly's Basin. The total length of line is 30 miles 41 chains."

" RULING GRADIENT.—With the exception of the first 3½ miles, where there is a gradient of 1 in 30, there will be no grade steeper than 1 in 40. This 3½ miles may be regarded as a separate section, as it is a down gradient from the Mine, and all the heavy loading will be down hill to the Smelters. The 1 in 40 grades are not compensated on the sharp curves, but there is a sufficient margin of level and easier grades to permit of this being done on a permanent survey."

" CURVES.—The sharpest curves on the line are of three chains radius. There are many places where the deeper cuttings could be reduced 50 per cent. by substituting curves of two chains radius for these, and sufficient information has been obtained to enable us to take out an estimate on this basis should it be required."

" GENERAL FEATURES, &c.—With the exception of four miles at the Lyell end, the formation is mostly schist and freestone on the hills, and peat resting on gravel beds on the button-grass plains. The rock will be easily excavated, and the cuttings will stand with nearly vertical sides. On the plains it will be merely necessary to strip the overlying peat to an average depth of about fifteen inches, when the gravel substratum will afford a very suitable formation. The slopes on the sidelong country run from 10 to 45 deg., averaging about 25 deg. In the conglomerate

country the surface is very hard, but the local experience of this
rock shows that it is usually a surface covering or boulder, under-
laid by schist or soft shales. Abundance of the best quartz
gravel for ballast is obtainable through the plain country in the
direct course of the line, and also at Kelly's Basin and Lyell.
Timber suitable for construction purposes will also be obtained in
sufficient quantities in the course of clearing the formation
widths."

" INTERFERENCE WITH PROPERTY, &c.—With the exception of
a few mineral leases near Lyell, the route is entirely through
unoccupied and unalienated Crown Lands—barren and unsuitable
for settlement."

"Owing to the dense scrub in the hill country, we were unable
to get the best results on a mere trial survey, and feel confident
that a much improved section could be obtained on a permanent
survey."

MOUNT LYELL BLOCKS MINING CO. N.L.

This property adjoins the Eastern boundary of the North
Mount Lyell Leases ; it consists of seven blocks of 10 acres each
(see locality map at title page).

Up to time of going to Press I regret having been able to
procure any particulars regarding this property, and am therefore
reluctantly compelled to give it this mere passing notice in my
pages.

I am aware however that it is extremely well situated on one
of the main big spurs leading (from the Southern side) of Lyell's
Main Peak—and that there are some good 'Shows' upon it
such as to make it quite worthy of being vigorously prospected
and opened up.

In the *Mount Lyell Standard* of October 23rd 1897 the
following notice or report was published concerning this property :

LYELL BLOCKS.

"Since my last visit to the Blocks property, the prospects have changed for the better. After 650 ft. of driving, the decomposed schist country gave way to quartzite, carrying a little copper and iron pyrites, which has continued for the last 30 ft. driven. The tunnel is now being extended towards the shaft sunk on the eastern boundary of the property, and close to the North Lyell ore-body. This will be continued until the shaft is reached, which will be in about 150 ft., and crosscuts will also be driven in order to prove the formation now being passed through. At the point where the shaft will be reached, the tunnel will be 180 ft. from the surface, so that it will test the property at a fair depth."

And in the same Journal's columns the following additional information about this property was published on December 8th 1897 :

THE LYELL BLOCKS.

"The tunnel has been driven 740 ft., and after passing through some 30 ft. of quartzite, the face is now in soft schist, with indications of the quartzite making again. The drive will reach the eastern boundary of the North Lyell property in another 70 ft. The manager hopes to encounter something good before reaching that point."

This is practically all the information I am in the position at the moment to contribute about this property.

THE LYELL THARSIS MINING COMPANY.

Capital **£12,000** in 24,000 shares of 10s. each : 20,000 shares
paid up to 5s. ; 4,000 shares held in reserve.

Secretary : JOHN POTTS, 39 Queen Street, Melbourne.

Area of property 21 acres.

This may now be considered as one of the "Crotty Group" of
Mines—that gentleman having quite recently entirely reorganised
its Board and practically taken over the future development and
control of this extremely promising Mine into his own personal
supervision.

The following is an Extract from Mr. Lawson's Book before
mentioned. (August, 1896).

THE LYELL THARSIS MINING COMPANY.

No Liability.

Capital **£12,000** in 24,000 shares of 10s. each ; 20,000 shares
issued paid up to 5s. ; 4,000 shares held in reserve.

Legal Manager : JOHN POTTS, 39 Queen Street, Melbourne.

" Area of 21 acres, adjoining the North Mount Lyell. The latest
work done at this mine consists of a tunnel which has been driven
125 ft., meeting, in its first course of 59 ft., schist, intermixed with
erubescite veins, the remaining 66 ft. being in lode formation
carrying copper and iron pyrites, with a little gold and silver."

" Men are at the present time engaged in cutting a chamber at
a point nearly in the centre of the lode formation, with a view to
sinking a winze in order to test its value at a depth."

The following paragraph was published in *The Lyell Standard* on December 15th last :—

"The Lyell Tharsis Mining Company (no liability) has been registered in Melbourne. The number of shares in the Company, which has been refloated for the purpose of raising capital, is 150,000, of £1 each. The number of shares subscribed for is 120,000. The original number of shares was 24,000, of 10s. each, 4,000 of which were held in reserve."

The following Article also appeared in *The Mount Lyell Standard* of October 23rd 1897 :—

THE LYELL THARSIS.

"This property continues to open up in a satisfactory manner. The Nos. 1 and 2 tunnels have been driven 60 ft. through payable ore, and they are to be continued as long as the ore body maintains a bulk value of not less than 4 per cent. of copper."

"In order to further prove the length of the formation, a trench has been started a considerable distance south of and parallel with the tunnel workings, and operations have disclosed the existence of the ore body identical in character with that through which the tunnels are being driven."

"It is intended to prove the length of the ore-body by trenching still further south and towards the huge outcrop which occurs above the present workings. At a depth of 110 ft. below the upper tunnels an ore-body 86 ft. in width has been passed through, which, though of a low-grade character, may pay if worked on a large scale."

"The winze which is being sunk in No. 2 tunnel is down 24 ft. Right from the start the ore has slightly improved. The bottom of the winze now shows a large amount of copper pyrites."

"Should the part of the ore-body being sunk through in the winze maintain its present average value, the Lyell Tharsis will have a deposit of ore of a proved extent of 166 ft. in length and 110 ft. in depth, while as to its width, that has been determined to be 60 ft."

D

MOUNT LYELL CONSOLS MINING COMPANY N.L.

This Mine is another of Mr. James Crotty's Northern 'Group' of Mines, and probably one of the best of them since he is 'nursing' it so patiently; and I expect to see it take a prominent place in the market at no remote date.

It is not only splendidly situated (on the south boundary of the "North Lyell" Mine)—but it has been most successfully exploited—to an extent indeed unknown to any but a very few of the "inner ring" of Lyell Mining men.

In Lieut.-Colonel Boddam's, R.E., admirable plans (issued and published under Mr. James Crotty's personal directions and now to be seen in scores of City Mining Offices) the position of the large and valuable lode and hematite outcrops surmounting same in this property are most clearly shown—and I strongly commend these plans to my reader's attention.

It will be seen that one at least of the "North Lyell" lodes almost certainly traverses the whole length of this property (and also almost the whole length of Mr. J. Crotty's adjoining property at its south boundary "The Mount Lyell Extended.")

The following notice of this Mine appeared in *The Mount Lyell Standard* on November 13th 1897 :

LYELL CONSOLS.

" Useful work of a preliminary nature is being done on this property, which has excellent prospects. As soon as the necessary initial operations have been completed, the Mine will be vigorously developed, and the extent of the ore-body determined. **It is fairly certain that the same class of rich ore as that which the North Lyell** has in such immense quantities will be met with. The process of opening up these Copper Mines in rugged country is unavoidably slow and tedious ; but when they have been opened up they are worth all the trouble."

MOUNT LYELL EXTENDED MINING COMPANY N.L.

Capital £150,000; in £1 shares; 60,000 paid up to 20s.; 60,000 paid up to 15s.; and 30,000 shares held in Reserve.

This Mine is also one of Mr. Crotty's Northern Group—and is situated at and adjoins (on its south boundary) the north boundary of the most northerly Mount Lyell Mining and Railway Company's block of 10 acres; and on its northern boundary this "Extended" Mine adjoins the south boundary of "The Mount Lyell Consols Company's" southern block.

In Colonel Boddam's plan it is shown that the chain of hematite iron outcrops—which so markedly act as 'Beacons' to either the 'adjoining' or the 'underlying' copper Pyrites deposits and lode (or—as in the "North" Mine—with Erubiscite or peacock ore &c. &c.) are numerous throughout the entire length of this "Extended" property; and thereby stamps it as being of very great value indeed.

This Mine is being opened up by an Eastern tunnel and drives &c. and the ore exposed in these workings is stated to be both abundant in quantity and of high grade; and the fact that Mr. Crotty regards this property as quite one of his best 'Shows' in this northern part of the field, entitles it to be considered as a Mine which will—when opened up—justify the high estimation in which he holds it.

THE PRINCE LYELL MINING COMPANY N.L.

Capital £100,000 in 100,000 shares of £1 each; of which 50,000 are fully paid up.

Legal Manager: G. A. LAWSON, MELBOURNE.

The following Extracts are from Mr. Lawson's Book before referred to: (August 1896):—

This property consists of 25 acres, situated west of West Mount Lyell Company, and about half-a-mile north-west of Crotty's Celebrated Iron Blow. The ground is being opened up by a tunnel and shaft, both along the strike of the lode, from 7 to 8 chains apart, by means of which the lode is being tested in two different places, and there is little doubt that if in the course of time the present shaft and tunnel be connected by means of the 100 feet level, the mining of good ore will be proportionately increased.

The Tunnel is now in about 350 feet, giving 150 feet of backs; 500 feet of backs will be available by going deeper. Latest assays from the face gave :—Gold, 5 dwts.; silver, 36 ozs.; lead, 1 per cent.; copper, 47 per cent. per ton.

The following extract is taken from a special report on the Prince Lyell Mine made by Mr. G. Thureau, F.G.S. (late Government Geologist and Chief Inspector of Mines, Tasmania) :—

"I have not found it difficult to place this property in the front rank of Mines located on the Mount Lyell Mineral Field, for the following reasons, viz.: There appears to be an almost unlimited supply of ore, which improves in depth; there exists unusual facilities for working these mines for many years to come by means of inexpensive adits, as no shafts are at all requisite, the construction of the road recommended is only a trivial matter of expense, in which the adjoining companies north might be invited to join. And as to permanency of the formations at Mount Lyell generally, that has been proved beyond a doubt by the very extensive and deep workings of the parent Company, and the diamond drill bore at the Central Company's lease, which intersected the lode or cap at 615 feet and has since gone through solid lode matter. Now if it is borne in mind that taking your Company's lode at say 165 feet wide, a stope 6 feet high and one foot 'in' along its strike would at least produce from 50 to 60 tons of marketable ore, if worked by stoping or open cut, and as your Company holds over twenty chains of that lode, the calculations as to the necessary working expenses and profits is a very easy matter to compute.

"The Mining Manager has just reported an important discovery made on the property. The lode has been struck in trenching, showing a splendid body of copper pyrites with erubiscite copper going strongly under foot. Assays give 32 per cent. copper. The discovery is of great significance, showing that large and valuable ore deposits exist outside the Lyell Mine.

"Having had many years' experience in mining of all description, I can recommend this property with every confidence' as to its eventual success to capitalists as a safe, enduring, and lucrative investment."

Recently the want of a sufficiently large working Capital necessitated a suspension of the prospecting and development work at this promising little Mine, but the following extract from *The Mount Lyell Standard* of October 23rd 1897 shows that work had been resumed at this mine :

PRINCE LYELL.

The management of this Company have wisely decided to continue prospecting their property, and operations were resumed this week by starting a further extension of the tunnel. Six men have been put on, working three shifts, and the tunnel has been extended a total distance of about 250 feet. The pyrites which I mentioned in last week's notes as being met with is still present, and is showing in a band of ore from 3 to 6 inches thick across the face, about three feet above the floor of the drive. This is where the principal bulk of the stuff occurs, but both above and below this band of ore the pyrites is freely distributed through the formation. This band of ore is not nearly so valuable as was at first expected. Where the solid pyrites occurs it is chiefly associated with clean white quartz, but the main portion of the face of the drive is in silicious schist, and assays from this stuff have returned 2½ per cent. of copper. The tunnel will be continued for a considerable distance further—in fact, as far as it is possible to do so. It will be decidedly advisable to cross-cut in both directions from the drive.

. . . . . .

Extract from *The Lyell Mining Standard* October 16th 1897 :

THE PRINCE LYELL, CROWN LYELL AND QUEEN LYELL.

PROPOSED ABSORPTION.

The proposal of a London syndicate to take over the Prince Lyell, Crown Lyell and Queen Lyell properties is rather a strange one, seeing that the leases are not adjoining one another, but are some distance apart in very rugged country, the intervening spaces being held by other companies or individuals. Can it be that the London syndicate do not know the respective positions of the properties? The *Standard* has no desire to throw cold water on the project, but it must be said that the reason for one company acquiring these three properties is hard to find. No doubt they are all good prospecting shows, and deserve to be vigorously and intelligently exploited, and if they were contiguous it would be a reasonable proposition to bring them under one management for the purpose of systematically developing them. It is to be hoped that the London syndicate are aware of all the facts, for it is undesirable in the best interests of West Coast mining, that English capital should be invested here under any misapprehension. Money is urgently needed just now to carry on prospecting work on the three properties, and it is believed that it can be spent on any one of them with very fair prospects of success. The capital that has hitherto been sunk in the Prince has, as recent developments prove, certainly not been expended to very good advantage ; but the tunnel might have been driven at a lower level with better success. At the Crown, owing to lack of sufficient means, the management have only been tinkering for some time past.

If the London syndicate are aware of the locations of the properties, well and good ; but why not acquire the Prince and amalgamate with the North Prince and the West Lyell Extended,

which adjoin one another and could be worked very well under
one management? The Crown should be absorbed by the North
Lyell or the Tasman Lyell, both of which are progressive companies : and the Queen would also be better in the hands of the
Tasman shareholders.

It is proposed, if the London syndicate do take over the
properties, to have a working capital of £150,000. That should
be ample to thoroughly test the value of each property, and it is
to be hoped that, if the deal eventuate, the money will be
judiciously spent. It is folly to attempt the exploration of Lyell
mines on small capital.

THE CENTRAL MOUNT LYELL MINING COMPANY.

No Liability.

Capital £30,000 in 100,000 shares of 6s. each. (See tabulated
list on map.)

Area of block : ten acres.

Situation : adjoining west boundary of Mount Lyell Mining
and Railway Company's Block No. 14, and also adjoining
south boundary of West Lyell Company's Southern Block,
No. 166.

In my Book (August 1896) I referred to the fact that the
" Diamond Drill " was being used in this Mine for the purpose
of ascertaining at what depth the underlay and hanging wall of
the " Big " Mine's ore deposit or lode would be found. It is
probably at the present date known to most of my readers that
in two out of the three Drill holes the pyrites body was struck at
depths averaging about 650 ft. from the surface, and that the
drill was then driven some 150 to 170 ft. into the solid pyrites
before work was discontinued. The pyrites ore found in the
' cores ' did not assay high in either copper, gold or silver, but of
course it is easy enough to conceive that a small 2-in. Drill hole

could (and certainly **would**) find plenty of " poor " or low grade " two inch " places even in the wonderfully ' uniform ' pyrites in the ' Big ' Mine itself.

And it should certainly be borne in mind in esti-mating the probable value of the "Central" Mine that this western or hanging wall part of the Lode has been found to contain quite as high grade ore as the average ore of the eastern or centre portions of the deposit ; and that extremely rich ore was not long since reported as having been struck on this western part of the Lode in the Big Mine itself (*vide* Mr. Powell's official reports &c.)

This property will probably be 'amalgamated' with one or other of its neighbours ; and with sufficient capital at command may quite well yield its quota of payable ore at a depth.

The following reprinted matter is extracted from my Book— as a "refresher" to such of my readers who may not have recol-lected same :—

MINE MANAGER'S REPORT.

To the Chairman and Directors of

The Central Mount Lyell Mining Company

(No Liability).

Gentlemen,

Herewith I beg to submit my Half-yearly Report for your consideration.

The following works have been carried out, viz. : Building two huts for the accommodation of the men when sinking the shaft takes place : also manager's dwelling-house, 24 feet by 12 feet, the latter being built of iron. Also erection of diamond drill over the No. 1 shaft, which had been sunk to a depth of 73 feet during the previous half-year. A further depth of 717 feet was bored, making a total of 790 feet. The object of the bore was to prove the underlay of the Mount Lyell lode into our property, which is now an established fact. The hanging wall was reached at a depth of 568 feet from the surface, and at 46 feet from the hanging wall the pyrites became solid, continuing so for a depth of 101 feet. At this point the bore passed into schist, which I believe to be an intrusion into the lode. This continued without change for a distance of 68 feet, when the footwall was reached, consisting of a band of ironstone 10 feet wide, and then we passed into conglomerate. I may here mention that a few assays were made from the first few feet of pyrites after it became solid. The results of these have been forwarded to the office. Further assays will be made from the lower portions of the lode, which will give a better idea as to its value. I will endeavour to have these completed in time for your meeting.*

We have now proved beyond all doubt that this extensive ore body penetrates our property, and although we have not succeeded in striking rich ore, it must be remembered that, on account of the extremely small diameter of the bore as compared with that of a shaft, the chances were greatly against our doing so, as we may have passed by a rich deposit, missing it by perhaps a few inches. We must not lose sight of the fact that rich and poor ore exists in the Mount Lyell lode, and as that body maintains the same general character throughout its entire extent, there seems no reason to doubt that rich portions will also be found in our property.

As the Directors have decided to further test the value of the mine by means of the drill, I have now shifted the plant **three chains farther north.** Here I anticipate reaching the lode at a much less depth.

Recently rich ore has been discovered going south in the Mount Lyell Mine. It will be advisable for us to test this portion of our property with the drill to ascertain at what depth the ore will be met with in this part of our mine, and as the pyrites rises with the hill, it should be met with at a reasonable depth, and you would be better able to determine which would be the best place to put down a main shaft.

I am, Gentlemen, yours respectfully,

P. M. BALSTRUP, *Mine Manager.*

16th January, 1896.

P.S.—The second diamond-drill hole is stated to have struck the lode at 646 feet ; and third diamond-drill hole is now down some 400 feet at a point near the S.E. corner of the block. †

* ASSAYS OF THREE SAMPLES FOR CENTRAL MOUNT LYELL MINING COMPANY.

	Gold.	Silver.	Copper.
No. 1.—Pyrites and Schist from footwall	1 dwt.	3 oz. 0 dwt.	1·14 per cent.
2.—Pyrites and Fahl Ore, 660 feet	½ „	3 „ 13 „	3·38 „
3.—Pyrites, 678 feet - - - -	½ „	3 „ 0 „	2·73 „

F. R. POWELL, *Assayer.*

KING LYELL G. M. Co. N.L.

I regret having to class this property—for the moment— amongst the " Silent Mines "—for the same reasons I have before referred to—i.e. I have been unable to procure authentic late data in time to include same in this Booklet ; I am however aware that in addition to the usual "alluvial" deposits which this Company has been—in a somewhat desultory fashion, ' working' for their contents in native copper—pyrites—gold &c. a defined and apparently valuable 'lode' has been exploited in this ground and is now being opened up As to its extent &c. I am not now able to speak, but I have seen the reports thereon and have every reason to believe them, and also that this new discovery will prove to be an important and valuable find for the King Lyell's southern and adjoining neighbour " The Mount Lyell Proprietary Limited " northern blocks (late Minerals Block of 40 acres).

Apart altogether however from even this most promising new " find " in the " King " and " Proprietary " properties I regard

NOTE.—† Lode afterwards struck at 600 feet and was driven into a further depth of 136 feet.—M. R.

the existence of the known large area of its as yet unworked
Alluvial deposits as being quite well worth systematic "hydrau-
licing" with a sufficiently powerful plant of "Giant" Nozzles &c.;
and I also believe that this Alluvial deposit extends south of the
"King" over quite a large part of the 40 acres of the "northern
blocks" of "The Proprietary's" Mines.

THE MOUNT LYELL PROPRIETARY COMPANY, LIMITED.

By reference to my small locality map (facing title page)
it will be seen that in addition to having become the owners of
the 40 acres of Mineral Lease lands lately purchased by the
Melbourne "Minerals Company" from myself and my late
partner (Mr. A. Kelly), The Proprietary Company also possess
the largest consolidated area of mining lands on the Lyell field;
and that included therein are the principal "conglomerate" and
Alluvial deposits of Mount Owen's giant slopes and shoulders.

Within some of these very promising "conglomerate" areas
it is reported—by well known local experts—that a number of
hematite-iron outcrops have been discovered and more or less
prospected; and that there are immense possibilities before
"The Mount Lyell Proprietary Company" both in consequence
thereof, and also in view of their owning a sufficient area of
mining ground to permit of "splitting" into several mining
"Pups" or subsidiary companies if justifiable 'finds' are
developed therein.

In respect to their 40 acres—(northern blocks) which are the
"Big" Mines nearest eastern neighbour—I may say that at the
time my then partner and myself "pegged out" and took up this
mining lease in August 1891—I was influenced to do so mainly
because I found in about the centre of the block—and following
and forming the 'bed' of a small creek (or tributary of the
Linda Creek) an iron outcrop of a lode—of "hematite"—and—

to all appearances—of precisely similar character to the "Blow" and other "outcrops" in the vicinity thereof.

It would now appear—in view of the Experts' reports referred to—that it was a decidedly fortunate and good stroke of mining business that we did take up this ground; and I am naturally well pleased to find that it has now been acquired by so well capitalised and powerful a Company as "The Proprietary."

The following Extracts from this Company's Prospectus will probably suffice to afford my readers all the most essential and salient particulars as to this Company's property and prospects :

Extracts from the Prospectus of

THE MOUNT LYELL PROPRIETARY MINES, LIMITED.

(Incorporated under the Companies' Acts, 1862 to 1893.)

CAPITAL - - - - - - £500,000
In Shares of £1 each.

ISSUE OF 400,000 SHARES AT PAR.

Payable 5 - on Application, 5 - on Allotment, 5 - February 25th next, and 5 - March 25th next, or the whole amount can be paid up in full on Allotment.

Of this Issue £100,000 in Cash or Shares is for Working Capital. Leaving 100,000 Shares in Reserve for Additional Working Capital.

Directors.

SINCLAIR MACLEAY (Director THE IVANHOE GOLD CORPORATION, LIMITED), 157 Winchester House, E.C.

D. F. CARDINALL (Director and Trustee for the Debenture Holders of the Manchester Brewery Company Limited), 18 Cromwell Road, Brighton.

EDMOND KASTOR, 7 Rue Meyerbeer, Paris.

GERARD WELMAN, late Government Secretary, Selangor, Straits Settlements.

ALBERT T. WRIGHT (Messrs. WRIGHT, BECKET & Co.), Water Street, Liverpool.

The Vendors have the right to nominate another Director after Allotment.

Tasmanian Advisory Board.

The HONBLE. NICHOLAS J. BROWN, Member House Assembly, Member Executive Council, and late Minister of Mines and Public Works for the Colony of Tasmania, Hobart.

The HONBLE. N. E. LEWIS, Member Executive Council, and late Attorney General for the Colony of Tasmania, Hobart.

A. G. D. BERNACCHI, J.P., Hobart.

Consulting Engineer.

MACNAMARA RUSSELL, Member Institute of Civil Engineers, 57A Park Street, Grosvenor Square, W.

Bankers.

THE LONDON AND MIDLAND BANK, LIMITED, 52 Cornhill, E.C., and Branches.

IN AUSTRALIA—BANK OF AUSTRALASIA

Brokers.

G. H. & A. M. JAY, 17 Old Broad Street, and Stock Exchange, London.

PIXTON & COPPOCK, 12 Half Moon Street, Manchester.

JAMES KIRKWOOD & SON, 62 Buchanan Street, Glasgow.

R. G. LAWS, 12 Mount Steuart Square, Cardiff.

Solicitors to the Company.

ASHURST, MORRIS, CRISP & Co., 17 Throgmorton Avenue, E.C.

Solicitors to the Vendors.

BLYTH, DUTTON, HARTLEY & BLYTH, 112 Gresham House, Old Broad Street, E.C.

Auditors.

CRAIG, GARDNER & Co., 41 Moorgate Street, E.C., and at Dublin and Belfast.

Secretary (pro tem.) and Offices.

F. A. HORNE, 8 Princes Street, Bank, E.C.

I reprint the following Extracts from the Reports published in this Company's Prospectus for the information of my readers :—

The 40 acre block.

Mr. G. Thureau, F.G.S., late Government Geologist and Chief Inspector of Mines, Tasmania, has made an exhaustive report on this block, which, as already stated, immediately adjoins the Parent Mine, from which the following extracts are taken :—

"Reports on five blocks of mineral land* containing about "thirty-nine (39) acres, more or less, situate in the county of "Montagu Mount Lyell, West Coast of Tasmania. The three "principal sections, viz., No. 893 M., 75/92 M. and 76 92/M., "immediately adjoin the Eastern boundaries of the famous Mount "Lyell Mining and Railway Company's Reward Claims, and that "boundary is about **one chain only** east of the celebrated

* NOTE : Mr. Thureau here refers to these 40 acres of " Mineral leased land " as " 5 blocks,"— these latter being in reality the 5 **gold-lease** blocks which merely over-ride this Mineral Lease—and which were applied for by and granted to the " Big " Mine's proprietors – M. R.

"'Iron Blow' (Crotty's) thereby occupying a most prominent "position on the Mount Lyell Mineral Field. The other sections, "numbered 31/93/M and 32 93 M. respectively, are located further "east, but adjoin the former.

. . . . . .

"In the year 1886 the Tasmanian Government instructed me, "their then Government Geologist, to proceed to and examine for "an exhaustive report the 'Linda *Goldfield,' its auriferous "resources, &c., &c. That report was presented to both Houses "of Parliament by his Excellency the Governor's command, in "October of that year: and on page 8 1 reported as follows: "' Besides gold, **copper,** chiefly found in its pure metallic state, "' occurs along a zone about nine (9) chains east of the original "' 'Iron Blow' (Crotty's). One Vein measures twelve inches and "' two others 2-3 inches wide. It occurs in quartzite, embedded "' in a kind of hard clay of brown colour, and appears to account "' for the lumps of pure copper found in Messrs. Watson's Claim, "' North Mount Lyell, weighing from 1½ lbs. to 6 lbs. each. "' Sometimes from 1½ lbs. to 2 lbs. of pure native copper can be "' washed in a dish.' "

"**There cannot be the slightest doubt that very** "**valuable copper deposits traverse right across the** "**first three of your sections as mentioned above in** "**this report** . . . Subsequently other lodes and cupriferous "deposits have been found, still further east, some of which trend "in the direction of your leases. I also discovered in the same "line of country, north (blocks), this copper to be associated with "' Fahl Ore,' exhibiting a high value in silver, so that I have no "doubt of your property being traversed by these valuable "metalliferous deposits, as a glance at the chart will show. . . .

"**Conclusion.**—It is well known that the report to Govern- "ment, referred to above, has been, during a period of nearly "ten (10) years of energetic mining and prospecting on this "valuable Mineral Field, proved correct, in every particular. ". . . . As to the position of your sections, I have in a "concise manner described same in portions of this report. **I can**

"only say in addition, from my intimate acquaintance
"of the same, that it can scarcely be surpassed.

"I have, therefore, no hesitation in recommending your
"property to investors as a safe mining venture ; with careful
"management based on experience and worked on economical
"lines, the final outcome will be, I am certain, lucrative and
"profitable for years to come."

Mr. RUSSELL states as follows respecting this portion of the
Property :—

. . . . "Mr. A. Kelly and myself decided to 'peg out' and
"make application for this 40 acre Mineral Lease at the time of
"our visit to Lyell . . . and in addition to the fact that our
"Western boundary adjoined the Eastern boundary of the 'Big'
"Mine's main Blocks, No. 13, 14, and 15, and were, therefore, in
"the closest proximity to 'Crotty's Iron Blow." I observe that
"in the bed of a small creek, running through about the
"centre portion of our block, there was exposed for some
"chains in length a very promising-looking hematite
"and gossany iron outcrop, bearing approximately about
"N. 10 W., and trending, therefore, into the 'King Lyell'
"Block, in which latter a portion of the Linda Creek Alluvial
"Gold Workings was still yielding native copper, pyrites, etc., in
"conjunction with more or less 'payable' gold. . . . . From
"what I have recently heard through trustworthy sources, this
"40 acre Mineral block of ours must almost certainly prove to
"be of much greater value than, personally, I ventured to assign
"to it in 1891 when we took it up. . . . . It is known to
"your Company, I believe, that I still hold a fairly large interest
"as a Shareholder in 'The Minerals Company,' from whom your
"Company has acquired your title to this 40 acre Mineral Lease."

———

Subsequently to the 40 acres block being pegged out as a mineral
lease by Messrs. Russell and Kelly in 1891, it was re-pegged by
the Mount Lyell Mining and Railway Company as Gold Leases in
four blocks of 10 acres each ; but, under the laws of the colony, no

gold leaseholders are entitled to mine for gold upon any portion
of the said blocks upon which this Company shall be *bonâ-fide*
conducting mining operations. The grant of the Gold Mining
Leases to the Mount Lyell Mining and Railway Company will
not, therefore, in any way interfere with this Company in
carrying on their mining operations on the 40 acres block under
the mineral leases.

The 527 acres.

Mr. GEORGE D. GIBSON, M.A., INSTITUTE, M.E., has made
exhaustive reports upon the Company's territory comprised in
the above-mentioned 527 acres, and which is almost the immediate
extension south-easterly from the Great Mount Lyell Mine. In
the course of these reports Mr. Gibson states that there are
**three distinct lodes traversing the property without
regard to the great lode of the Mount Lyell Mine, but
that in his opinion the latter lode also in its natural
course south-easterly runs directly through the Company's property.** Upon this point the following is an extract
from his report :

" As already stated I am decidedly of opinion that the track
"of the Mount Lyell lode will be found to cross the north-
"eastern angle of the property. . . . My opinion is arrived
"at from a study of the surface indications. On the eastern, or
" footwall, side of the great lode there occurs a remarkable belt,
"or dyke of red jasparised sandstone totally unlike the prevail-
" ing rocks of the district. . . . . To the south-east after
" leaving the Lyell Hill it becomes covered with alluvial soil, but
" re-appears on the rising ground in the Rio Tinto. . . . The
" same unmistakeable formation can be followed climbing
" diagonally up the slope of Mount Owen and just below one of
" the highest peaks again makes an immense outcrop. Even at
" this great elevation—over 2,000 feet higher than the site of the
township and about 3,200 feet above sea level—hematite iron
"outcrops strongly and pyrites occurs sparingly, the mineral
" following closely the strike of the red sandstone. The appearance
" of hematite and never so little pyrites at such a height is a

" matter of vast importance and the occurrence helps greatly to
" remove any doubts that might hitherto have existed as to the
" continuance of the Mount Lyell fissure, **and I fully share**
" **the opinion held by Dr. Peters that the red belt of**
" **rock referred to undoubtedly marks the track of the**
" **Mount Lyell lode.** If, therefore, the footwall backing of
" the lode can be traced right from the Mount Lyell Mine to the
" north-east angle of this section, it must naturally follow that
" the body of the fissure lies within your Company's area; it
" also appears that the pyritic body underlying the ironstone
" capping pitches from the Mount Lyell Mine to the south-east,
" but again rises on the slopes of Mount Owen."

The following is also extracted from this prospectus:—

" On November 24th, 1897, a cable was received stating that
" a very valuable ore body had been struck on the King Lyell
" Mine, near the boundary of the 40 acre block owned by the
" Mount Lyell Proprietary Mines, Limited. A cable was sub-
" sequently sent to the Legal Manager of the Great Southern
" Mount Lyell Mining Syndicate, at Melbourne, asking for
" further particulars regarding this important discovery of
" valuable ore, and whether it extended into this Company's
" property. To this a reply was received on the 12th instant, as
" follows:—' **The lode runs north and south through our**
" **property. Width not yet determined. Have driven**
" **in 5 feet. An assay of this sample gave 37 per cent.**
" **of copper solid underfoot.'** "

THE COPPER MINES OF MOUNT LYELL WEST LIMITED.

CAPITAL - - - - £400,000

In 400,000 Shares of £1 each.

Viz.: 195,000 fully paid up Shares of £1 each to Vendors and a "First Issue" of 35,000 Shares of £1 each to Public (also now fully "paid up" for Working Capital &c.); and 170,000 Shares are held in Reserve for future issue for producing further Working Capital when required.

Directors.

CHARLES McCULLOCH (Chairman African Gold Recovery Company, Limited), 7 India Buildings, Liverpool.

CHARLES J. BUCKLAND, F.G.S. (Director Associated Australasian Miners, Limited), Suffolk House, Cannon Street, E.C.

HERBERT PALMER (Director Colenbranders' Matabeleland Development Company, Limited), 4 Drapers' Gardens, E.C.

A. E. FERNS, Kirby House, Heaton Chapel, Stockport.

MACNAMARA RUSSELL, M.Inst.C.E., 3 Great Winchester Street, London.

Bankers.

THE MANCHESTER AND LIVERPOOL DISTRICT BANKING COMPANY, LIMITED, 75 Cornhill, London, E.C.
MANCHESTER, LIVERPOOL AND BRANCHES.

Brokers.

Messrs. CUTCLIFFE, LEY & McCULLOCH, 7 Adam's Court, and Stock Exchange, London, E.C.

Messrs. MARSLAND & CHEW, Leinster Chambers, 4 St. Ann's Square, and Stock Exchange, Manchester.

Messrs. OUTRAM & HAMILTON, 82 West Nile Street, and Stock Exchange, Glasgow.

Solicitors.

Messrs. HEPBURN, SON & CUTCLIFFE, Bird-in-Hand Court, Cheapside, E.C.

Auditors.

Messrs. F. J. SEARLE, SMITH & CO., Chartered Accountants, 4 Sun Court, Cornhill, E.C.

Secretary and Offices.

H. A. H. RUSSELL, 3 Great Winchester Street, E.C.

Agent in Tasmania.

J. B. HICKSON, Elizabeth Street, Hobart.

This valuable property is at last becoming pretty well known to the Public thanks to the successful flotation of the present Company in London a few weeks since : and the opinions I expressed so freely and confidently about it (in my Book in August, 1896), will during the next twelve months—and possibly even much sooner—be within 'test' point.

With a present subscribed Capital of £30,000 (less what has been lately expended in sending to the Mine a first instalment of Machinery sufficient to enable the new Manager—who has been appointed—to push on vigorously with the three principal 'prospecting' workings &c.) I have every confidence that this property will amply justify and reward both those who have supplied the present Capital and also the more numerous body of general speculators who have lately made so strong and satisfactory a ' market ' for its shares.

And I shall venture to take this opportunity of reminding my co-Shareholders that in the 'prospecting' works which have been conducted (in more or less precarious fashion for want of sufficient Capital to provide plant and machinery &c.) at this Mine for the past four years or so—some most important and encouraging discoveries and developments were made.

In the " Kelly " Shaft, for instance, which is situated within " an arm's length," so to speak, of the " Big " Mines most northerly (and north-westerly) tunnel and other workings, and which are but some 200 feet from our Eastern boundary, highly important and promising lode matter was met with. This Shaft is 156 feet in depth ; and at the 150-ft. level a southerly drive was put in for a total distance of about 375 feet. **In this Shaft massive blocks of Hematite iron, and iron and copper pyrites in big lumps and boulders were struck at a depth of about 35 feet from surface ; and this class of lode matter continued right down to the 135-ft. level, at which point the conglomerate footwall—or usual eastern wall—of the lode was struck.**

The Shaft was sunk into this conglomerate down to the 156-ft. level ; and at the 150-ft. level a southerly drive was then put in. The first few feet were driven in conglomerate; then came some 20 feet of "schists"; and then followed **the conglomerate footwall** of lode; this was penetrated for some 30 feet, and **a solid body of hematite iron of the very finest and densest character was then met and driven through.** The drive was then extended into the schist hanging wall—(rather uselessly as is now evident)—for some 200 feet or so.

Instructions have now been given to resume work in this "Kelly" shaft, and Steam drills and other items of a preliminary prospecting plant have been shipped to the Mine. It is intended to open out upon the hematite iron lode formation met with in this Shaft and to follow same down, &c.; and personally I have the greatest possible confidence that we shall strike the solid pyrites ore body itself in our proposed new Western drives winzes and cross-cuts. Expectations which may be considered strengthened and rendered almost a certainty—in my opinion—by the fact that although the lode was said to be considerably "disturbed" in the No. 4 northern level of the Big Mine's workings it was there unquestionably "**trending**" direct into our ground *vide* Mr. R. Powell's reports &c. Whilst in the No. 2 Bench, or "open-cut," the **outcrop** of the "Big" Mine's pyrites lode is actually seen on the surface at a distance of only 100 feet from the S.W. corner peg of our most Southern block (*vide* Mr. Muir's report &c.).

It should be distinctly remembered by my readers—or those of them who are "interested" as shareholders in this property—that the "Kelly" shaft is situated so close to the Big Mine's Main Workings and to the developed pyrites lode—as to make it almost an impossibility that the lode itself is not in our ground at this point; whether—however—it will prove to extend and continue as "one and the same lode" say to our next nearest proved "lode formation" distant some 600 feet northerly in the now well known "Razorback" Spur—of course we have no sufficient evidence yet before us.

In theory however this to my thinking will most probably

prove to be the case and in fact I believe it is almost certain that it will be found to be so.

In our "Razorback" Tunnel driven beneath a huge mass of conglomerate "outcrop" and which forms it is now generally believed the exposed Eastern **footwall** of the Lode ; and with 30 to 40 ft. of solid 'Chert' already met with in the Tunnel in its usual and most characteristic juxtaposition with the "Conglomerate," I feel the utmost confidence that when we have extended this Tunnel right through the "Chert" mass we shall be rewarded by meeting with the main copper pyrites ore body itself in its customary contact with these "Cherts" and "Conglomerates" thus forming its usual "footwall."

Our new Company is quite alive to the immense significance and importance of the indications I have referred to—and are losing no time in giving practical effect to their views.

For myself—having been originally induced to 'peg-out' and apply for in August 1891 (in conjunction with my then partner) these then "forfeited" gold lease blocks which to-day form the new Company's property—in consequence of my finding both hematite iron and copper pyrites in the two shallow surface prospecting Shafts at the "Razorback" and in the northern block at "Russell's" lode as it has since been christened—it cannot be surprising to my readers that I have such exceeding faith now in the future of this Mine of ours.

The "Russell" lode—situated some 1,900 ft. northerly of the "Razorback" Tunnel—bids fair to be also a very satisfactory and additional "feather in my cap"—or witness to the correctness of the estimate I formed of its possible value so far back as August 1891—when as I have said I first made its acquaintance.

During the past year or two quite a large number of competent experts—local as well as from Tharsis—Tinto &c. &c., have examined this 'Show' and have pronounced the lode and outcrop &c. to be of immense prospective importance ; and assuredly it **is** so ; for from the surface down to the 75 ft. level of the "Russell" underlay 'Eastern,' or foot-wall, Shaft solid hematite iron of finest quality has been sank through and splendid bodies solid of copper pyrites have also been met with assaying from 3 per cent. to 14 per cent. of copper;—Sulphides of Copper—and at the 60–75 ft. levels **native copper** extending right across the whole width of the Shaft—have been passed through.

No time is now being lost in resuming work at this lode— its potential possibilities being quite well appreciated by my colleagues as well as by myself.

Whether or not Mr. J. Crotty's theory—(as referred to by him in the report published in our Prospectus) will prove to be correct is premature to say; certainly a glance at Lieutenant-Colonel Boddam's excellent plans of the Northern Group of "Crotty Mines" indicates almost conclusively **that our " Kelly Shaft " and " Razorback " lode will be found to be the continuation of the " North," " Consols," and " Extended," " Main Lode "**; in which case this powerful and well-defined "Russell" lode with its N.N.W. strike and 40 feet of width at the surface &c. &c. must either be a 'loop' lode to the "main lode" aforesaid, or it must be a distinct "Western lode" or ore body or "deposit"; as to which—Time and Mine developements alone will disclose to us the real facts.

In conclusion I shall only point out to my readers that not in one of the many of the other mining properties at Lyell is there a Mine which occupies or can possibly occupy the "priority of situation or place" such as our property enjoys; whilst its indications, proved prospects, and mining advantages are of very great magnitude, and especially when the large area and the size and great length of its lode formations are duly considered.

And assuredly my co-shareholders are not called upon to exercise any very great stretch of imagination or faith when they are invited to believe (as I strongly and earnestly here 'invite' them to do) that we have unquestionably got the continuation of the " Big" Mine's Main ore deposit itself in our southern block; besides various other "good things" in each of our more northerly sections.

THE SOUTH MOUNT LYELL MINING CO. LIMITED.

Capital £600,000 in 300,000 shares of £2 each.

London Offices of the Company—138 Leadenhall Street, E.C.

Managing Director—D. J. MACKAY, Esq.

Area : nine and a half blocks = ninety-five acres.

Situation : adjoining in part the Central Mount Company's south boundary and Mount Lyell Mining and Railway Company's blocks (*vide* map).

This immense property is also one of the "Crotty Group" of Mines—it being the most southern one thereof. So much is now known to the Public generally about this Mine—thanks to Mr. J. Crotty's connection with it—and to the satisfactory practice which the various administrators of his Companies adopt of keeping the Shareholders and Public advised almost day by day of all matters pertaining to the progress of mining developments &c.—that I deem it superfluous for me to express either my opinions or to give "data" about this property.

In my Book (August 1896) I published very full "Reports" &c. from the various local experts and other scientific men who bore testimony to its great prospective value.

What this latter will actually prove to be is now fairly well on its way to being disclosed at no very remote date—seeing that a vigorous working policy characterises all Mr. J. Crotty's various undertakings. But even he cannot hurry on such matters as necessarily involve "making haste slowly"—such as erection of heavy machinery sinking shafts &c. &c.

That it is however a "potential" property of the first class amongst the Lyell Group there can be no question ; and as a "Crotty" favourite should—I venture to consider—command public confidence.

THE GREAT SOUTH LYELL MINING Co. Ld.

This is a recently formed "Consolidation" Company which has acquired a very large area of Mineral-lease property situated at the southern end of the Lyell field; *vide* my small locality Map.

I have however no information whatever about this property I regret to say—at time of going to Press.

THE SOUTH MOUNT LYELL CONSOLS LIMITED.

This is the title—I am informed—of another new Company in embryo which has acquired the large area of mineral lands situated at and adjoining the western boundary of Mr. J. Crotty's South Mount Lyell property.

At time of going to Press no particulars or details of this new Company or its prospects are available to me.

The following is extracted from Mr. Lawson's book :

GREAT SOUTHERN MOUNT LYELL MINING SYNDICATE.

No Liability.

Capital £4,000 in 4,000 shares of £1 each.

Legal Manager : ALFRED PFAFF, 409 Collins Street, Melbourne.

Area of 60 acres, situated in the southern portion of the Mount Lyell mineral field, on the south-western slope of one of the spurs of Mount Owen, and lies in a south-easterly direction from the principal workings of the famous Mount Lyell Mine,

which are about three-quarters of a mile distant, although
the north-west angle of the section abuts with the southern
extremity of that Company's extensive property. It also adjoins
the Rio Tinto Company's ground on the south, and the South
Mount Lyell Extended on the east. Mr. Geo. D. Gibson, M.A.
Inst. M.E., in his report, states, the property is traversed by
three distinct lodes, apart from any reference to the Great
Mount Lyell lode, the track of which in his opinion crosses
the north-east angle of the block, and if this is the case the
property would command the advantage of its western underlie to
the very fullest extent. No. 1 lode has not as yet been traced
through the section, but is seen outcropping just beyond the
boundary, near the south-eastern angle, at the head of a gully
which cuts across the southern portion of the block. It is
impossible to say much about the nature of the lode at present,
beyond that the gossan, which forms the capping, is of a most
promising character, and from its appearance indicates the
existence of copper at a depth. This proposition is further
confirmed by the existence of native copper in the stream beds,
and the manager informs me that on washing even the surface
gravel both gold and copper can be obtained. No. 2 lode. Nearly
all the work done in the past has been confined to exposing this
lode in a series of openings and trenches along the line of strike,
and at present preparations are being made for driving a tunnel
from a point in the south-east portion of the claim in a northerly
direction, bearing 20 degrees to the eastward, to intersect the
same at a depth of 225 feet, which he estimates should be attained
in a distance of about 160 feet. Ascending the hill northwards,
on the line of the tunnel, an opening has been made in which the
capping of the lode is first exposed, consisting of silicious ironstone
of promising appearance.

Following along westward, I inspected the several trenches, in
almost all of which there is to be found hematite iron, in association with schist and more silicious ironstone.

In none, however, is the width of the lode shown until a point
is reached some 200 or 250 yards westward from the line of tunnel,

where a costean, commencing at the bottom of the valley, has been run up the hill in a northerly direction for a distance of about 250 yards.

This long trench reveals a wide extent of lode material fully 300 feet across, and composed, as in the other openings, of mineralised schist and silicious ironstone, with veins and patches of hematite iron throughout. No. 3 Lode is seen cropping out strongly in the north-eastern portion of the section, striking north-westerly, and running in the direction of the South Mount Lyell Mine, from which it may be assumed that it corresponds with the lode cut in that company's No. 1 Tunnel. Mr. Gibson is also decidedly of opinion that the track of the Mount Lyell lode will be found to cross the north-eastern angle of the property, and considers the property is one of great prospective value.

The following item from the *Mount Lyell Standard* of December 19th may also interest some of my readers :—

"The property of the Great Southern Mount Lyell Mining Syndicate, No Liability, has been successfully floated in London into a Company of 500,000 shares of £1 each, on terms which are very favourable to this Company, whose shareholders will receive 125,000 shares of £1 each, fully paid up, less a commission of ten per cent. A further 175,000 will be held in reserve for future issue if required, and the balance allowed to the London promoters, on their placing £25,000 to the credit of the Company for working capital. Signor A. D. G. Bernacchi, who left Melbourne a few months ago, has carried the flotation through very successfully. The English Company will (says the *Mining Standard*) be styled "The Mount Lyell Proprietary Company Limited."

THE "SILENT MINES" OF THE MOUNT LYELL FIELD.

It will be probably noticed that whilst on pages No. 23 and 24 I gave a list of some 29 mines which are held by various Registered Companies (mostly, be it noted, Melbourne "No Liability" companies—having very nominal amounts of " Working Capital ") that I have only given particulars or details of quite a select few companies or properties.

That this is so is due entirely to the fact of my having been unable to 'collate' data from any reliable source whatever (up to the time of this booklet going to Press) as to their position financially or as to their " workings "—if any—of what I will term these " Silent mines." And it must not therefore be supposed that because I have nothing to say about them that I either ignore them, or believe them to be valueless ; as regards some of them I feel sure they will not remain "silent" much longer, for they undoubtedly possess prospects such as will warrant Capitalists in taking up the properties and providing sufficiently large " Working Capital " &c.

THE GEOLOGICAL CHARACTERISTICS
OF THE LYELL FIELD.

———

Every mining field, and every isolated ' Great Mine ' has its own characteristic—or individuality—no matter however apparently close is its analogy in one or more general features to some other near or distant mining celebrity, or ' mining field.'

I have been frequently asked by many readers of my ' Book,' and by others—to explain what is the meaning—in plain non-technical language, of Dr. Peter's and other professional men's references in their reports (upon the Lyell Mine itself—and the field generally) to " Aqueous deposits " of " lode " or metalliferous ores in pyritic and other forms.

I lay no claim to being either a geological authority—or qualified to teach even the rudiments of the vast and most interesting science of Geology—or of its twin sister Mineralogy ; both however have formed pretty constantly during many years past a very favorite study with myself as a ' mining man ' ; and I now therefore venture to include in these closing pages a couple of rough sketches indicating graphically—if not very accurately or scientifically—the theoretical principles which these presumed Aqueous-deposits " seem to warrant us in thinking were the

AUTHORS' "THEORETICAL" SKETCH VIEWS OF THE LYELL MINERAL
DEPOSITS "BEFORE" AND "AFTER" THEIR ASSUMED "UPHEAVAL" "EPOCH."

① "BEFORE UPHEAVAL"

SECTION THRO' CHAIN OF HILLS AND LAKES.

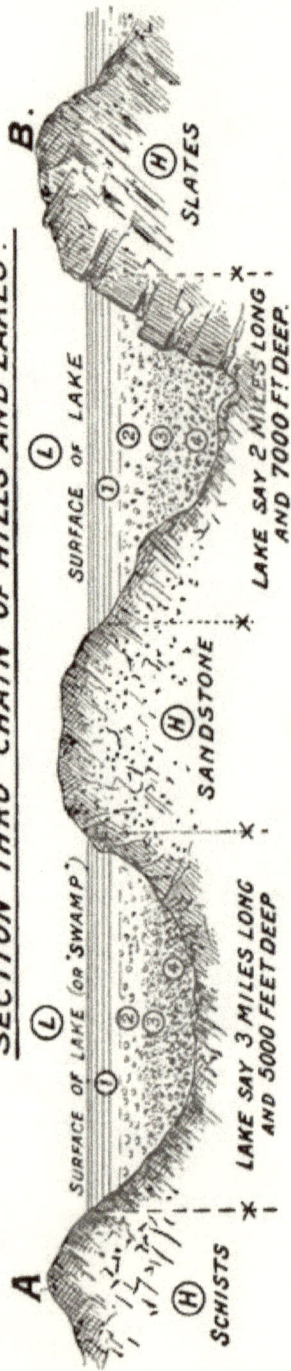

A.

B.

SLATES

SCHISTS

SANDSTONE

(H)

(H)

(H)

(L)

(L)

SURFACE OF LAKE (OR SWAMP).

SURFACE OF LAKE.

LAKE SAY 3 MILES LONG AND 5000 FEET DEEP.

LAKE SAY 2 MILES LONG AND 7000 FT. DEEP.

① ② ③ ④

REFERENCE:

(H) The enclosing Hills with   (L) Lakes.

① Water; ② Gravels, ③ Fine Sands (metalliferous - from Matrix
Sources etc.) ④ Coarse Gravels.

M.R.
JAN. 1898.

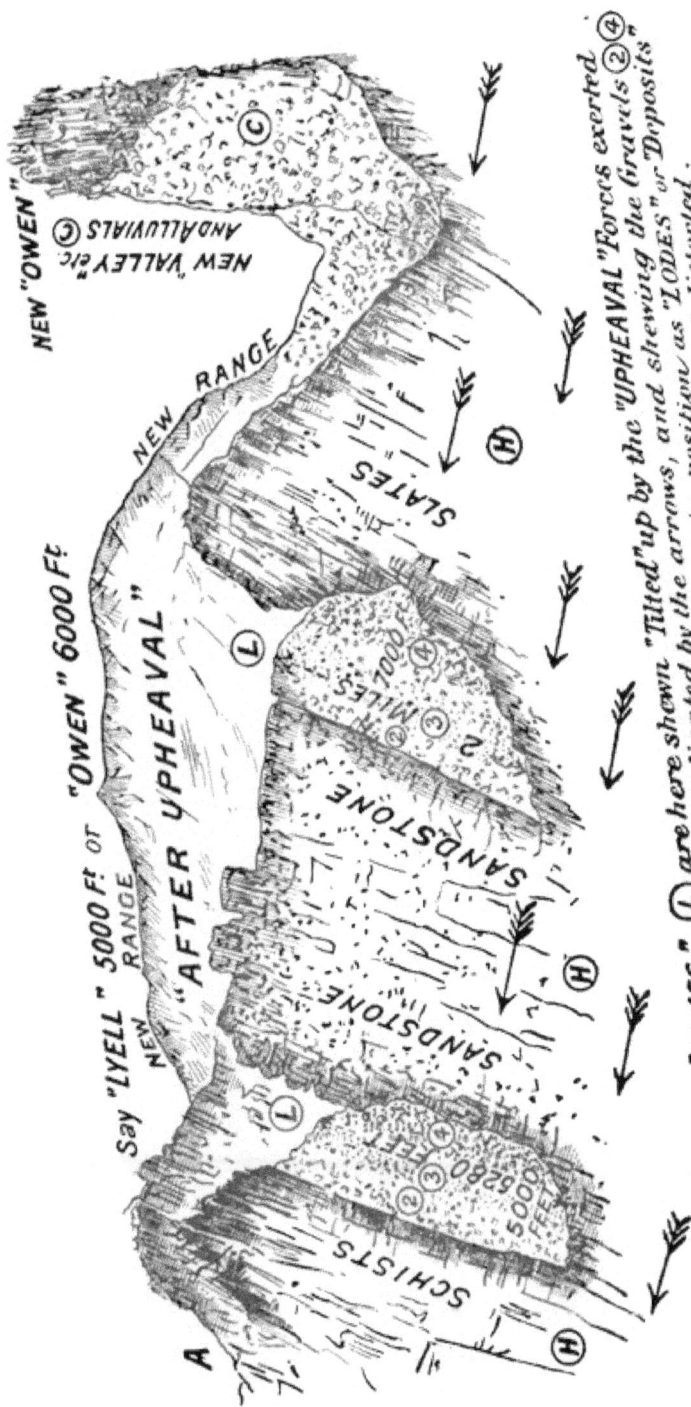

NEW "OWEN"

C

NEW "VALLEY" &c. C
AND ALLUVIALS

NEW RANGE

Say "LYELL" 5000 Ft or "OWEN" 6000 Ft
NEW RANGE

"AFTER UPHEAVAL"

L

L

SLATES

H

2 MILES
3
4
7000

SANDSTONE

SANDSTONE

H

L

5280 FEET
2
3
4
5000 FEET

SCHISTS

H

A

NOTE: "LAKES" (L) are here shown "Tilted" up by the "UPHEAVAL" "Forces exerted (2) (4)
in the direction indicated by the arrows, and shewing the Gravels (2) (4)
and metalliferous fine Sands (3) now in position as "LODES" or "Deposits"
in their enclosing matrix "country" upheaved and distorted.

M. R.
Jan. 1898.

probable principal factors in their original formation as infiltrated and eroded gravels and metalliferous fine-sands of varying grades or degrees of richness according as their deriving matrix-rocks held rich or poor metalliferous veins pockets lodes etc.

In Sketch No. 1 I have assumed that at the pre-eruptive epoch of this Lyell Field, with its now huge "Lyell" and "Owen" Peaks and Mountains of 5,000, to 6,000 feet and their precipitous valleys and gorges—there existed an elevated Lake district (similar indeed to that which at some ten or twelve miles north of our Lyell Peaks of to-day is to be seen—with its Lakes "Spicer," "Dora," "Beatrice," "Rollason," &c.) ; and we may suppose that in the course of many centuries the "beds" of these mountain-fed Lakes were by the aid of scores of creeks or rivers, and some possibly hidden or subterranean springs, also gradually 'fed' with these sub-aqueous deposits of varying thicknesses, and of still more varied metalliferous contents : some of which—such as the particles of various forms of iron ores, and the copper-yielding 'combinations,' with their more or less closely allied and precious associates of Gold and Silver,—were subjected to some of the infinitely numerous, and in some cases quite inexplicable chemical 'actions' and 're-actions' in Nature's own marvellous Laboratory.

And whilst this 'deposit' and these natural chemical processes and 'actions' went silently and placidly along throughout the long centuries, doubtless there were many periods of disturbance of the Earth's crust of more or less violence : and doubtless also these Lacustrine Deposits, or accumulations of miles-wide and of possibly even also miles-deep Lake-Beds, suffered 'distortion,' both lateral and vertical, by such volcanic outbursts and their effect upon the Earth's crust within the area affected thereby.

And on "one fine day"—(a "fine day" indeed for Lyell's Shareholders—*per exemple*—who were not however even to be born for many centuries later) there 'came along' one of those terrible terrestrial convulsions whereof this Earth of ours is so full of over-abundant evidences—and which in its all over-

powering and resistless force tore through such child-like impedi-
menta as a few miles more or less of the Earth's crust" and thereby
paying off, possibly, some old grudge or family quarrel with that
placid Lacustrine elevated plateau, with its mile-long chains of
Lakes and Swamps and with their hundreds of miles of rivers and
creeks running more or less rapidly through the 'Matrix'—
metalliferous rocks; and by their 'aqueous' eroding "action"
bearing off the metalliferous particles thus released; and which
were then persistently—and perpetually—"deposited"—(throughout those long centuries of Time)—into these lake "beds" and
Swamps of various lengths depths and breadths.

And—behold it: on that "fine day" aforesaid, the peaceful
Lacustrine panoramic scene has suddenly vanished: gone—for
ever and ever; its placid Lake Waters and miles of Ancient
Rivers sent tearing away in mad fear—and wild career—to find
new 'beds' whereon to repose, and wherein haply to found
new and still newer "deposits"!.

And as to those old—those hundreds of centuries old—
Lacustrine deposits—trace their fate and future—as depicted in
my Sketch No. 2;—and we perceive that they are no longer
reposing in their 'beds' of pristine Lacustrine bed-making, but
have been "tilted" bodily either bolt upright or at more or less
sharp vertical angles of 'dip' and 'underlay,' and with more or less
distorted and twisted "trends" or "bearings"; and that hence-
forward and for ever they shall know no more their erstwhile
peaceful 'beds' and Lacustrine bed-fellows . . . . but shall
lie in abject subjection at the feet of their all-conquering
disturbers of the peace—those Earth-crust burglarious Intruders
from below—and who now rear their proud "Lyell" and
"Owen" heads, shoulders, and broad backs, thousands of feet
into the Heavens and look down—in Supremest indifference—
upon those poor Lacustrine beds or 'deposits' now here and there
showing in their mangled, tip-tilted imprisonment amongst their
rocky enclosing 'foot' and 'hanging' walls—here an 'outcrop,'
and there a stretch of 'exposed' gravels or "conglomerates,"

or of metalliferous "deposits" which have gone through those mysteries and 'Actions' of Nature's Laboratory aforesaid—and are changed now magically into—

— Well : say into ore deposits, iron outcrops, and 'cherts' and cemented gravels and sands, and "conglomerates" of every degree of size and colour and form, **but everywhere** still sticking closely—as such long centuries--old friends should do— to their very oldest of friends, and in fact most intimate "blood-relations" those metalliferous "deposits" aforesaid.

To my readers let me now say *vale—vale—valete*.

———

F I N I S.